People, Science and Technology

People, Science and Technology:

A Guide to Advanced Industrial Society

Charles Boyle
Senior Lecturer in Science and Society
Trent Polytechnic

Peter Wheale
Senior Lecturer in Management and Business Studies
Oxford Polytechnic

Brian Surgess
Course Tutor
The City University Business School

BARNES & NOBLE BOOKS
Totowa, New Jersey

First published in the USA 1984 by
BARNES & NOBLE BOOKS
81 ADAMS DRIVE
TOTOWA, NEW JERSEY 07512

© Boyle, Wheale and Sturgess, 1984

Library of Congress Cataloging in Publication Data

Boyle, C.
  Man, science, and technology.

  Bibliography: p.
  Includes index.
  1. Technology – Social aspects.   2. Science – Social aspects.   I.
Sturgess, Brian T.   II. Wheale, P.R.
III. Title
T14.5.B69   1984   303.4'83   83-24368
ISBN 0-389-20455-2

Printed in Great Britain

To
Michael, Katy
Emma and Caroline
From
Peter

Contents

PART II   Key Issues in Advanced Industrial Society

PART III   Questions of Control

x   *Contents*

Preface

The aim of this book is to provide an introductory account of the social relations of science and technology, that is, of the ways in which science, technology and society interrelate. It has been written for science and engineering students mainly, but should also be useful to students of social studies, and of interest to the general reader. No specialist knowledge in either the humanities or the natural sciences is required to understand the arguments presented.

Formerly, many teachers of the science and engineering disciplines regarded their sole duty to students as being the transmission of proficiency in technical knowledge and appropriate associated skills. Recently, however, there has been an increasing recognition of the need to discuss the social implications of advances in scientific knowledge and of technical innovations. The 1980 Finniston Report on the future of British engineering, for example, emphasised the importance to future engineers of an adequate appreciation of the broader business context in which their work will be done. Other reports have stressed the need for scientists and engineers to be able to communicate more effectively with non-technical colleagues, and to contribute more effectively to policy-making. If they are to improve their relatively low status in Britain, where at present, in comparison with other industrial nations, they are under-represented in positions of authority, scientists and engineers must show an informed awareness of broader issues that is not easy to acquire with a narrowly based education.

Resulting from these considerations, and from an increased sense of the social responsibility of the scientist and technologist, has been the broadening of scientific and technical education in institutions of higher and further education and in schools. This book is a contribution to this process, and we hope that as well as providing a general background to the study of the social relations of science and technology, it will encourage students to pursue more deeply specialised topics of particular interest to them.

We wish to express our gratitude to Ruth McNally, who gave us invaluable editorial help and advice, and to Mrs Audrey Bode and Diane Coates who kindly typed the manuscript.

1 Introduction

CHARLES BOYLE AND PETER WHEALE

1 Structure of the book

It is difficult to imagine a discussion on the possible paths of development of the modern world, or on the potentialities of human life in the future, which does not almost immediately focus on science and technology and their role in social change. Nor is an analysis of the problems facing the world at present likely to mean much, if it does not involve some attempt to understand the complex interactions between science, technology and society. Science and technology, at many different levels, are intimately bound up with our hopes and our fears. They have given us standards of comfort and technical marvels unimagined formerly, and they promise more; but they pose threats too, graver in some ways than those faced in the past. Their products permeate our environments at home, at work and at leisure; and they invade our minds and shape our consciousness with the ideas they promote.

Science, technology and society studies draw on many disciplines. The chief concern of these studies is not with peripheral, abstract and academic issues, but with problems which are central to our lives. They are projected at us from our television screens and newspapers every day: shortages of food; suffering and ill-health; scarcity of energy resources; disruptive effects of new technologies and new modes of communication; and, above all, threats of devastation by war – these are concerns which demand attention and action. They are not the sort of problems that, if ignored, will go away. They are diverse, but they have one thing in common: science and technology are deeply implicated in all of them, both as part of their cause and part of the attempted 'cure'.

There is a need in any democratic and pluralist society for citizens to be well-informed, and hence capable of appreciating and discussing the issues at stake in their changing social environment. Snow

(1963) and others have pointed out that a gulf exists in industrial society between scientists and engineers on the one hand, and, on the other, those who by education or inclination are 'arts-orientated'. Scientists are often dismissive of the knowledge of non-scientists, and vice versa. We believe a wider study of the social relations of science and technology could lead to a bridging of the gulf between the 'two cultures', a better understanding between those who regard science as superior to all other knowledge, and those who see science and technology as superficial and vulgar.

Part I of this book provides a set of perspectives on science and technology from the viewpoints of history, philosophy, sociology, politics and economics. Part II applies these different perspectives to each of the following areas: food and agriculture; health and medicine; energy; military technology; scientific management and work; and telecommunications and the mass media. Part III returns to the general themes outlined in Part I, and looks in particular at the crucial problems of the regulation and control of science and technology.

The overall structure of the book is illustrated in Figure 1.1, and is further explained in sections 2 and 3 of this chapter. The book provides a general framework for thinking about the social relations of science and technology, and an introduction to a range of topics that seem to us to be of outstanding importance. References are given at the end of each chapter and for the reader who wishes to pursue further study of the topics discussed, a guide to further reading is provided at the end of the book.

2   Perspectives

The terms 'science', 'technology' and 'society' are broad, with many layers of meaning that have accumulated over the centuries (Williams, 1976). We define below the meanings we imply when these terms are used in this book. According to Bernal (1969), science can be taken

(1) as an institution; (2) as a method; (3) as a cumulative tradition of knowledge; (4) as a major factor in the development of production; and (5) as one of the most powerful influences moulding beliefs and attitudes to the universe and man

We shall deal with science in all these senses. By technology we

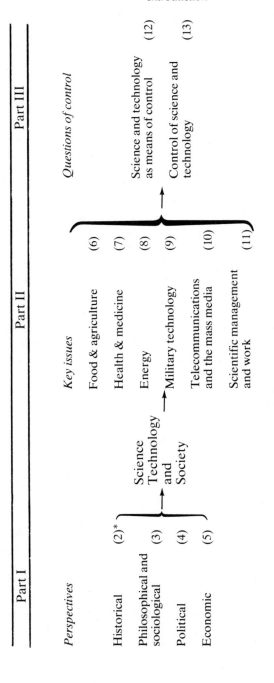

Figure 1.1: *Structure of the book*

* Numbers in brackets refer to chapters

Part I · Part II · Part III

Perspectives

Historical (2)*

Philosophical and sociological (3)

Political (4)

Economic (5)

Science
Technology
and
Society

Key issues

Food & agriculture (6)

Health & medicine (7)

Energy (8)

Military technology (9)

Telecommunications and the mass media (10)

Scientific management and work (11)

Questions of control

Science and technology as means of control (12)

Control of science and technology (13)

mean the knowledge, skills and equipment used for various purposes, including industrial production: it thus covers both 'software' and 'hardware'. The term society refers to the whole complex of relationships and institutions (legal, governmental, religious and cultural) within which people live.

How should we visualise the interactions between science, technology and society? Figure 1.2 shows three simple models. Figure 1.2 (a) suggests a simple linear development, with discoveries in science leading to technological applications which produce certain effects in society. Alternatively, it has been suggested that changes in society make demands on technology and these lead to scientific innovations. In support of the first we might think of basic pure research in electricity and magnetism in the nineteenth century leading to the study of electrons, the technological application of this research in the development of television, and the resulting effects on society of television in terms of information and news, entertainment, propaganda, advertising, and so on. In support of the second, it is argued that in the eighteenth century there was a clear economic need – a demand by society – for a new source of power. This led to the invention and gradual improvement of the steam engine. Not until this engine had reached near technical perfection was its scientific basis clearly understood and formulated in the new subject of thermodynamics.

To suggest that such simple linear developments tell the whole story is, however, naive. The reciprocal effects on society of steam technology were profound, whether in mines, factories, railways or ships. Clearly, in the first example, neither the science nor the technology behind television sprang from nothing; only because of financial and other support from society for the necessary research and development work did the technology of television come into being. Social forces too influenced the *form* the technology was to take.

Many examples spring to mind which lead us to reject these simple models and to replace them by Figure 1.2 (b) which takes account of a far greater number of possible interactions. Though this is a more satisfactory model, and has been used as a basis for analysing many case histories, we reject it too because of its treatment of science and technology as autonomous areas with clearly defined boundaries. Rather, they interpenetrate and overlap; one needs only to reflect on a few technical tools – the microscope, or

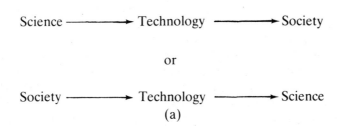

Science ——————→ Technology ——————→ Society

or

Society ——————→ Technology ——————→ Science

(a)

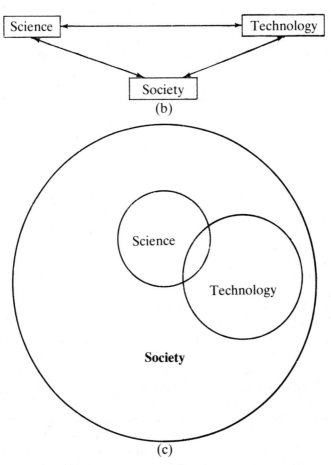

(b)

(c)

Figure 1.2: Models of Science, Technology and Society (a) linear model, (b) interaction model, (c) 'embedded' model

the telescope, say – which at various stages in history have transformed whole areas of science. These in turn have been transformed themselves by the fundamental theory science has brought to bear on them, and by the impetus it has given to the development of new forms, such as the electron microscope or the radio telescope. Further, science and technology cannot be considered to be external to society; they are, rather, embedded in society, and influenced by all sorts of political and cultural preferences and constraints which are expressed most clearly in terms of differential financial support for various fields, but also in other ways. At a philosophical or ideological level there have always been exchanges between scientists and other intellectuals, with attempts by many groups to use natural science and its jargon to legitimate political or ideological doctrines and social policies. Scientists are social beings as much as other groups and are subject to similar social and economic pressures. It would be most extraordinary if such pressures did not leave their marks in some ways within science itself. Technology, even more than science, is intimately shaped by socio-economic pressures and is permeated by the culture in which it is created and used.

It is for these reasons that we choose the model depicted by Figure 1.2 (c) which we have called the 'embedded' model, showing science and technology, as it were, immersed in the ocean of society, which soaks through them as through a sponge. Throughout we use the insights of social studies to provide perspectives on science and technology. The historian, for example, will study social change over time; the sociologist will be interested in the scientific community, and in the ways in which it resembles or differs from other communities, and in which it produces socially agreed knowledge; the economist will study the level of financial support given by governments and private companies to research and development, how these resources are allocated among the various branches of science, and how innovation affects the production of material wealth; the political analyst will consider, among other things, government control over science, and the political and military implications of technological developments; while the philosopher will critically examine the nature of scientific knowledge and its claims to truth. By applying all these different perspectives to areas of science and technology we believe a better understanding of advanced industrial society emerges.

3 Key Issues in Modern Society

We begin with a brief discussion of the concept of post-industrial society, then examine briefly the issues selected for analysis in Part II, and finally mention the main themes to be found in the book.

Industrial society, as it has developed since the early nineteenth century, has been sharply distinguished by historians from the basically agrarian, village-centred society which existed before the Industrial Revolution. Some sociologists, and in particular the American Daniel Bell, have claimed that now, in the late twentieth century, we are on the threshold of a similar great social revolution which will bring into being a post-industrial society as different from the society of the present as the present is from the pre-industrial era. Different writers have given different names to the new society, and emphasise different aspects of it; some welcome it, others condemn it; but there is broad agreement about its key features.

Industrialisation brought about a movement of workers from the primary sector (agriculture and extractive industries) to the secondary sector (manufacturing industries). In recent years there has been a shift from the secondary to the tertiary sector (service industries) of the industrial economies. The USA and the UK, for example, have over half of the work-force employed in such areas as finance, education, health and retailing. It is suggested that this has led to an increase in white-collar rather than blue-collar work, and to the increasing professionalisation of jobs. An important feature of a post-industrial society is the central role of knowledge, particularly theoretical technical knowledge of the sort that is embodied in computer systems. According to Bell (1973): 'Industrial society is the coordination of machines and men for the production of goods. Post-industrial society is organised around knowledge for the purpose of social control and the directing of innovation and change.' This view holds that universities and research foundations will be among the most influential institutions of the future, and great power will be held by technical experts. The post-industrial society will also be a post-scarcity society, and will have, because of very high levels of productivity, an abundance of goods of all sorts. It will be in effect a science-based welfare state, where the pursuit of profit will be replaced by the provision of services by professionals, and where the work ethic of industrial society will have been replaced by a new set of more relaxed values, appropriate to a life of leisure.

These ideas have been subjected to strong criticism. It is denied that we are experiencing any fundamentally new social revolution. Kumar (1978) for example, writes that

> what are projected as novel patterns of development turn out on examination to be massive *continuities* within the basic system of the developing industrial society . . . the trends singled out by the post-industrial theorists are extrapolations, intensifications, and clarifications of tendencies which were apparent from the very birth of industrialism.

The growth of the service sector, the shift to professionalised work, and the increase in importance of theoretical knowledge have all been occurring since the early nineteenth century. It is also argued that the statistics Bell uses in some of his arguments are suspect. For example, numbers of professionals show an increase because more people now *call* themselves professionals or are classified as such, without doing truly professional work. On the contrary, some suggest that many professional fields are being undermined and jobs deskilled by the application of computer technology. Furthermore, is it appropriate to use, as Bell does, figures of expenditure on research and development, or on education, as measures of theoretical knowledge or of an educated society? These and similar challenges lead to a much less optimistic picture than the one Bell presents. His critics maintain that, far from living in a welfare state, we are living in a warfare state to judge from the scale of our military efforts. Competition and profit-seeking are as common and as intense as ever; and the idea of a post-scarcity society is a very bad joke indeed for the undernourished millions of the Third World. Although, as a term referring to the next few decades or even the present, the phrase 'post-industrial society' has crept into general use without many of the overtones given to it by Bell, we prefer the term 'advanced industrial society' to underline continuous links with the past.

There is considerable overlap between the areas discussed by writers on post-industrial society and those areas we have chosen to deal with in Part II, in which science and technology are so closely involved. We have focused on matters we believe are important to our future – food, energy, medicine and health, war, communications and work. All these areas are closely related to the economic base of society. They loom large too in the flood of futures studies of the last twenty years, studies outlining world futures from a variety of standpoints, whose predictions range from the technologically miraculous to the ecologically catastrophic.

In this book we examine a range of different questions involving a broad spread of major problem areas related to both the physical and biological sciences. Three important themes run through this book. One, mentioned already, is the nature of the connections of science, technology and society. A second is the theme of control, both of science and technology by society, and of society by its science and technology. The third relates to the future and deals with the crucial question of which technological, social and economic choices are open to us. Related to these three themes are two key questions (see below) which the reader is invited to consider throughout the book.

Francis Bacon in *New Atlantis* (1624) wrote: 'The end of our foundation is the knowledge of causes, and secret motions of things; and the enlarging of the bounds of human empire, to the effecting of all things possible.' These were the aims of an imaginary community of scholars, conjured up in a seventeenth-century science fiction blueprint of a research laboratory, with workers engaged in a wide variety of practical projects to improve the human lot. The Royal Society took its inspiration largely from Bacon's vision. He was only one of a long line of optimistic philosophers who saw science and technology as the means of liberating the human race from drudgery, famine, disease, natural disasters and war. Since Bacon's day science and technology have progressed beyond even his wildest dreams.

The key questions are: Why has the sort of human liberation Bacon envisaged not been achieved? And what can we do to bring it about?

References

Bacon, F. (1624), *The New Atlantis,* ed. Gough, A.B. (Oxford: Oxford University Press, 1924).

Bell, D. (1973), *The Coming of Post-Industrial Society: A Venture in Social Forecasting* (New York: Basic Books).

Bernal, J.D. (1969), *Science in History vol.*1 (Harmondsworth:

Penguin).

Kumar, K. (1978), *Prophecy and Progress: The Sociology of Industrial and Post-Industrial Society* (Harmondsworth: Penguin).

Snow, C.P. (1963), *The Two Cultures and a Second Look* (Cambridge: Cambridge University Press).

Williams, R. (1976), *Keywords* (London: Fontana).

Part I
PERSPECTIVES ON SCIENCE,
TECHNOLOGY AND SOCIETY

2 Historical Background

CHARLES BOYLE

1 Introduction

The main aim of this chapter is to provide a very brief account of the main developments in science and technology since the early seventeenth century as a background essential for the understanding of modern industrial society. It is only, we believe, by looking at the past that we can try to make sense of the present, and form some idea of what the future may hold.

The history of science and technology is extremely complex; to select outstanding advances and identify underlying trends is inevitably to over-simplify. Nevertheless, we have considered for convenience three periods (treated in sections 2, 3 and 4 respectively) – the Scientific Revolution (seventeenth century); the Industrial Revolution (1750-1850); and the Emergence of 'Big' Science and Technology (1850 to the present). Tables 2.1, 2.2 and 2.3 provide brief summaries of some of the main scientific and technological developments since 1600, and are given to supplement the text, providing extra information in very condensed form.

Historians differ in their approaches to the study of the history of science and in the accounts they give of its development. There is a divergence of opinion, for example, between 'internalists' and 'externalists'. The internalist school holds that science, both in its content and rate of growth, is independent of social and political forces. Here, science is considered to have some internal logic of development of its own, and its practitioners, depicted as acting out of 'pure' intellectual curiosity, are seen as seekers after a truth that transcends social and personal interests. Many modern accounts of the history of science, however, are externalist to a greater or lesser extent, and reflect the belief that social and political factors play a profound role in the development of science and technology (See, for example, Basalla, 1968 and Bernal, 1969).

In accordance with our model, outlined in Figure 1.2 (c), of science, technology and society as interpenetrating entities, we adopt the externalist view. Furthermore, we do not look for any *one* single determining factor or cause underlying or ruling over the historical process. There are technological, political and economic determinists who see technology, politics or economics, respectively, as *the* basic concept, the independent variable, on which, as foundation, their historical explanations must finally rest. We feel that an historical explanation of any movement or set of events must be multidimensional, taking scientific, technological, economic, political, ideological and other factors all into account. It is this approach which colours the material selected and presented in the following sections, but it is to be noted that only a brief listing of some important developments in science and technology is given. A true externalist account, relating these developments to their social and economic context, would require much more space than is available here.

2　The Scientific Revolution

Most historians of science see the beginnings of modern ways of thinking in the great scientific revolution of the seventeenth century, which brought about in a small but increasingly influential minority a new view of the world, a new attitude to nature, the universe and man's place in it.

The 'scientific world-view' (as we shall call it, though recognising our anachronistic use of the word 'scientific' with reference to the seventeenth century) provided a framework for novel thoughts and concepts, a mode of illumination in the light of which many dogmas from the past were examined and rejected, and fresh teachings hammered out to take their place. It was not the result of a sudden, drastic mass conversion, but came rather from a whole chain of small, deliberate, piecemeal developments. Many people were involved in its articulation, and laid different emphasis on different aspects; the most famous figures were Bacon, Kepler, Galileo, Descartes and Newton.

The scientific revolution occurred against a turbulent social and political background. The first half of the seventeenth century saw the Thirty Years War in a Europe torn by religious discord, and the

Civil War in England. What remained of feudal institutions and privileges was increasingly threatened by the rise of a new predominantly Puritan merchant class, who subscribed to doctrines of 'possessive individualism' that emphasised liberty, property rights and toleration, and were tailor-made to foster the growth of early forms of capitalism. This group and the intellectuals who supported them (and in particular the French *philosophes* who were their spiritual heirs in the eighteenth century, the Age of Reason, of the Enlightenment) were sympathetic to science, and saw technology providing a vehicle for unlimited human progress. They were critical of the absolute monarchy of Louis XIV and his successors in France, but had more success in England, especially after the Glorious Revolution of 1688, which, like the earlier Civil War, had enhanced the power of Parliament and the middle classes relative to the monarchy and the aristocracy. Many writers have seen close links between Puritanism, capitalism and science during the emergence of all three as powerful social influences (Merton, 1970).

We shall look in more detail in Chapter 12 at philosophical aspects of the scientific world-view and at some of its effects, but limit ourselves here to a consideration of the great scientific achievements of the seventeenth century (see also Table 2.1). Dominating them was Newton's construction of the theory of gravitation, uniting celestial and terrestrial motions within a single framework in the boldest and most sweeping manner imaginable. This brought to a climax the work of Copernicus in the sixteenth century and its seventeenth-century development by Kepler and Galileo. Closely linked to the theory of gravitation, indeed essential to it, was the establishment of mechanics as an accurate quantitative science on sure foundations. Other fields in which revolutionary advances occurred were optics and hydraulics. In addition to analytical geometry, Descartes contributed his influential mechanical philosophy, with its deliberate separation of mind and matter and its emphasis on the clear methods of mathematics. Advances were made in chemistry and biology (in particular the theory of the circulation of the blood), but these were of less import than those in the mechanical sciences.

The study of optics, and of the properties of gases, has obvious connections with practical improvements – in telescopes and pumps, for example. Some historians have argued that work on the apparently 'purer' disciplines, such as astronomy, mathematics and

Table 2.1: *Important developments in science and technology (1600–1750)*

Date	Science	Technology	Society
1600	Kepler – planetary orbits Galileo – mechanics Descartes – mechanical philosophy, analytical geometry Harvey – circulation of the blood Classification of plants	Problems of navigation Use of telescope Bacon's view of research	Early colonists in America Persecution of witches Thirty Years War – religious discord in Europe. Growth of early capitalism.
1650	Rise of English science Boyle – gas law Alchemical experimentation Huygens – the pendulum and the wave theory of light Newton – calculus, gravitation, optics	Mercury barometer Improved microscopes Study of pumps and gases	English Civil War. Growth of ideas of liberty, toleration and possessive individualism Founding of Royal Society and Académie des Sciences Constitutional monarchy in England. Louis XIV in France
1700	. . . dominates English science Linnaean classification Stagnation in English science	Savery's steam pump Newcomen Steam Engine Growth of small-scale manufacture Improved iron and steel-making	Establishment of public banks. 'Classical' architecture and art English mercantilist colonial policies
1750	Studies in electrostatics		

mechanics, was not without its practical significance too. Such work was in fact strongly stimulated, if not initiated, by the desire to solve certain pressing problems of navigation and surveying, and was supported financially in such institutions as Gresham College, London, by rich merchants engaged in the expansion of international and colonial trade.

Bacon (1561 – 1626) often failed to recognise the importance of the best work that was being done in his time, and his contribution was not to the content of science. It lay rather in his vision of a large body of men earnestly and systematically experimenting on a whole range of problems, building up a new corpus of knowledge, and consciously seeking to exploit it for the benefit of mankind. Technology, however, developed relatively little in spite of the fact that most scientists, like Bacon, were utilitarian in outlook and desired strongly to apply their discoveries. Finally, we must mention the establishment of the first great scientific academies – the Accademia dei Lincei in Italy, the Royal Society in England, the Académie des Sciences in France, which encouraged correspondence and rivalry between scientists all over Europe, in place of the uncommunicative isolation and secrecy of the scholars of the sixteenth century (Mason, 1956).

3 The Industrial Revolution

The new ideas introduced by the scientists of the seventeenth century had little direct influence on the lives of most people, but this is certainly not true of the innovations of the period from 1750 to 1850 (see Table 2.2). The textile industry was the first to be transformed, by the use of water and steam to power new types of machinery, and by the organisation of work in factories, where vastly greater productivity could be obtained from a controlled and disciplined labour force than from the scattered family groups who had worked at home previously. This growth had a multiplying effect. As cotton sales shot up, so too did the production of coal and iron. A chemical industry developed to manufacture the acids and alkalis, bleaches and dyes necessary to treat the cotton. A machine-tool industry grew to provide for the increasing demand for machinery, especially during the second phase of the revolution, the period of expanding railway systems that spread out in great

Table 2.2: Important developments in science and technology (1750–1850)

Date	Science	Technology	Society
1750	Early debates on origin of the earth, fossils, the flood etc. Phlogiston theory in chemistry Scientific expeditions Science influenced by nature – philosophy Discovery of chlorine and oxygen Overthrow of phlogiston theory Foundations of modern chemistry	Birth of scientific engineering Agrarian revolution Growth of ideas of technical progress Hot air balloons	The Enlightenment The Age of Reason British conquest of India Science as liberatory force French *Encyclopédie* published American Independence
1800	Foundations of geology Foundations of electro-magnetism Principle of conservation of energy Theory of thermodynamics Theory of biological cells Lead in science shifts from France to Germany	Improved textile machinery Much improved steam engines Vaccination Rapid industrialisation of Britain – factories, machine tools, coal, textiles Growth of transport – canals, roads, railways, shipping Telegraph Growth of gas industry	Adam Smith's economics French Revolution: 'Liberty, Equality, Fraternity' The Romantic Movement Malthus on population Population increase and shift to cities German university system
1850	Advances in agricultural chemistry		

networks across the continents of the world, and, in conjunction with the electric telegraph, revolutionised communications. Improved agricultural techniques were introduced to feed the increasing population displaced to the towns from the countryside (Landes, 1972).

Progress in science too was spectacular on a variety of fronts. The 1780s saw a clear break with alchemy and theories such as the phlogiston theory of combustion. Modern chemistry started with the work of Lavoisier (1743 – 94) and others, and grew rapidly. After 1800 in physics, current electricity and its various effects were systematically investigated and mathematical frameworks constructed. The great unifying principle of the conservation of energy was hammered out, and the laws of thermodynamics formulated. During this period too the hazy and confused speculations on floods and fossils, inspired often by Biblical studies, gave way to solid theory as the new science of geology was established. But the scientific theory that most firmly seized the public imagination was in the related area of natural history – the theory of evolution, published by Darwin and Wallace in 1858. Though this theory upset Church leaders, when enunciated in a simplified and disorted form it provided a ready supply of concepts such as natural struggle and survival of the fittest, which were much to the liking of propagandists for free trade, cut-throat competition and exploitation, and imperialist-racist attitudes (Young, 1973).

As important as the actual discoveries and inventions were innovations in the organisation of science in the nineteenth century, and improvements in technical education. The coining of the word scientist in this period indicates the rise of the specialised professional and the corresponding decline of the 'natural philosopher', the amateur of independent means, who had shaped eighteenth-century science. Research teams were set up, establishing a tradition in the universities, first in Germany, but later in other European countries and the USA. Later still, applied research and development groups were to be employed directly in various industries. Before 1800 most technological innovations were craft-based and were made not by men of learning but by craftsmen. If science played a role in the earlier stages of the Industrial Revolution, it was the indirect one of contributing to an intellectual climate which encouraged the application of scientific methods to manufacturing. Industrial innovators were often on friendly terms with scientists and engineers, and were fellow members of influential societies in the English provincial towns, such as the Lunar Society of Birmingham. After 1850, however, many of the most important new developments – in chemicals and pharmaceuticals, in the gas and electricity industries – were science-based, and could not have been made by men with no theoretical knowledge of science (Hobsbawm, 1976). As the nineteenth century progressed, Germany became the foremost scientific nation, and England's overwhelming dominance of technology and manufacturing industry gradually declined as Europe and the USA industrialised.

The social and political changes associated with the Industrial Revolution are well known. The population of Great Britain increased from about 11 million in 1801 to nearly 21 million in 1851, and the growth of the large industrial towns and cities was even more dramatic. Population, wealth and power, formerly dispersed throughout the country, became concentrated in the cities. The new rich at first challenged traditional centres of influence, but later allied themselves to the landed gentry, and eventually through marriage merged with them. Wealth was produced on an unprecedented scale in expanding world markets in an atmosphere of *laissez-faire* free trade. Money and profits emerged as dominant social forces, dissolving some ties and obligations, strengthening others. Caught up in less enviable positions in the network of frantic exchange, of buying and selling commodities (whether labour power, or raw materials from the colonies, or manufactured goods from industrial nations) were the wage earners, the urban proletariat, a whole new class of men, women and children who, if fortunate enough to find work, put in long hours of inhuman toil in factories and lived in hovels (Engels, 1845).

Adam Smith (1723 – 90) and later Ricardo (1772 – 1823) wrote their economics treatises during this period, while the wage-earners had their champions in Marx (1818 – 83) and the other socialist economists. Some of the writings of Malthus (1766 – 1834) and J.S. Mill (1806 – 73) are forerunners of the modern 'limits to growth' debate, with Malthus forecasting society's incapacity to produce extra food as rapidly as extra children to eat it; while Mill looked forward more optimistically to a balanced 'stationary state' of society.

4 The Emergence of 'Big' Science and Technology

Perhaps the most striking characteristics of the history of science over the last century or so are the growth in its *content* (the extent of scientific and technological knowledge) and the change in the *context* (economic, social, ideological) within which scientific work is carried out. It has been transformed from the fringe activity of a few thousand enthusiasts to a huge, complex, industrialised undertaking, central to the concerns of the modern state, employing millions, and swallowing up in the advanced industrial countries 2 per cent or more of the gross national product (that is, the total value of all goods and services produced in the economy).

First, what were the main developments in scientific knowledge during the last hundred years? In the physical sciences there has been the astonishing revolution associated with the theory of relativity and quantum theory, providing a sharp break with the Newtonian mechanics and classical physics that, to the complacent scientists of the late nineteenth century, seemed to provide an unshakable framework for all future research in physics. In these theories, during the years 1900 – 30, were to be elaborated the concepts and principles that, together with some later techniques, underpin work in all the familiar fields of high technology today. Some of the traditions of pure research are still maintained in fundamental particle physics, using ever more powerful and costly machines to study the composition and structure of sub-nuclear units which, two generations ago, were considered indivisible and ultimate.

The biological sciences have seen continuing, and indeed accelerating, progress along the road to reductionism, that is the explanation of all life-forms and processes in terms of 'matter in motion' or atoms in fields of force. Increasing use has been made of the physical sciences, as the terms 'biochemistry', 'biophysics', and 'molecular biology' indicate. The reductionist emphasis is very clearly apparent, for example, in the history of cell theory. The concept of biological cells, developed in the 1830s, was influenced by the mystical and romantic German 'Nature-Philosophy' with its use of terms such as 'World Spirit' and 'vital force.' Improved microscopy led to notable advances in the 1870s and 1880s when chromosomes and cell divisions were studied. The growth of organic chemistry and interest in the workings of inheritance have

Table 2.3: Important developments in science and technology (1850 to the present)

Date	Science	Technology	Society
1850	Theory of evolution Organic chemistry Periodic table of the elements Cell division, heredity Electromagnetic theory Physical chemistry X-rays, radioactivity, electrons	Industrialisation of USA and Europe Internal combustion engine Antiseptic surgery, germ theory of disease, public health Important German chemical industry (dyes, explosives, etc.) Electric power Industrial research labs founded	Free trade – ethos of competition, survival of the fittest Writings of Marx American Civil War Franco-Prussian War Growth of trade unions Colonial imperialism Growth of monopolies
1900	Theory of relativity Development of biochemistry Quantum theory Lead in science shifts to USA Genetics	Introduction of electronics, radio Aeroplanes Poison gas in warfare Car industry, oil US technological dominance Industrialisation of Japan	World War I Russian Revolution Depression of 1930s Rise of Fascism and Nazism World War II Dominance of USA
1950	Nuclear physics New tools, eg. electron microscope, radio telescope, computers Science policy studies Molecular biology – genetic code High energy physics	Nuclear weapons, missiles Electronic communications Synthetics, plastics, nylon Artificial fertilisers TV, consumer goods Nuclear power Space technology and exploration High technology medicine – new drugs, transplants, etc.	East – West Cold War Growth of mass entertainment Increased spending on science Growth of military – industrial complex Independence for colonies – widening gap between rich and poor countries
1980	Extensions in cosmology Microelectronics	Computers, information technology Increasing automation	Environmental concern Energy crisis

produced descriptions of phenomena associated with cells that are more chemical and molecular, and less concerned with problems such as consciousness. The development of the electron microscope and X-ray analysis, and the discovery of the structure of DNA in more recent times have contributed powerfully to these strongly mechanistic tendencies.

The triumphs of technology in the twentieth century are well known. A few important developments are listed in Table 2.3. Casting a long shadow over them all are the nuclear stockpiles, with their threat of unparalleled destruction. Behind them stand other complex weapons, delivery and communications systems; in no area of human endeavour has technology been more assiduously and successfully applied than in the military field. The revolutions in transport and communications inaugurated in the nineteenth century have still not come to an end. Mass ownership of cars, dating from the 1920s in the USA, a development that seems so obvious now, was very largely unforeseen by the Victorians; but it is air travel and electronic telecommunications which have transformed international communications.

A new understanding of the structure of matter has meant that the range of materials at our disposal has increased dramatically, especially since the Second World War, which gave a great boost to technical innovation of all sorts. These new substances – artificial fibres such as nylon, plastic of various sorts, chemicals for use as pharmaceutical agents, fertilisers, pesticides, food additives – in many cases are derived from oil. World consumption of energy has increased ten times since 1900, oil consumption over a hundred times. It is not surprising that by the late 1970s seven of the world's top ten companies were oil multinationals. Food production has also greatly increased, yet one quarter of the world's population is malnourished. The medical record too is mixed – some diseases have been eradicated, but others have come into prominence. Two areas which have become highly significant in the last decade are biotechnology, using the insights and techniques of molecular biology, and information technology, drawing on computing, microelectronics, telecommunications and automation.

That science has grown in its scope and in the resources it consumes is obvious. Quantitative assessment of this growth is not easy, but Price (1963) has shown that since the seventeenth century the number of scientific journals published annually has increased exponentially, with a doubling period of some fifteen years. Numbers of practising scientists have no doubt increased similarly, and the sums of money spent on research and development even more rapidly, though it is only recently, when they have been levelling off, that there have been accurate estimates of these latter figures.

The traditional image of science as a quest for pure knowledge, and of scientists as isolated, dedicated and free seekers of truth, oblivious to the uses to which their discoveries can be put, is no longer valid (see p.30). For example, even in pure research, such as high energy physics, teams of tens of experimenters working together on highly expensive equipment are more typical than the solitary scientist. But, more important, applied or 'mission-oriented' research dominates science today, consuming most scientific expenditure, employing most personnel, deciding and assigning work to be done. In countries such as the USA and the UK, according to OECD figures, nearly half of all government spending on science and technology is in the military field (see Chapter 5); most of the rest is closely tied to industrial developments of one sort or another; and only a few per cent of the funds can be said to go to 'pure' research, even when this necessarily imprecise term is interpreted broadly. Increasingly, science has become 'industrialised' and it becomes more and more difficult to think of it as the disinterested, value-free pursuit of truth its practitioners once claimed it to be.

References

Basalla, G. (1968), *The Rise of Modern Science: Internal or External Factors?* (Lexington, Mass.: Heath).

Bernal, J.D. (1969), *Science in History,* vols 1 – 4 (Harmondsworth: Penguin).

Engels, F. (1845), *The Condition of the Working Class in England in 1844* (St Albans: Panther, 1969).

Hobsbawm, E. (1976), *Industry and Empire* (Harmondsworth: Penguin).

Landes, D.S. (1972), *The Unbound Prometheus: Technological Change and Industrial Development in Western Europe from 1750 to the Present* (London: Cambridge University Press).

Mason, S.F. (1956), *Main Currents of Scientific Thought* (London: Routledge & Kegan Paul).

Merton, R.K. (1970), *Science, Technology and Society in Seventeenth Century England* (New York: Harper Torch Books).

Price, D. de S. (1973), *Little Science, Big Science* (New York and London: Columbia University Press).

Young, R. (1973), 'The historiographic and ideological contexts of the nineteenth-century debate on man's place in nature', in Teich, M. and Young, R. (eds.), *Changing Perspectives in the History of Science* (London: Heinemann).

3 Philosophy and Sociology of Science

CHARLES BOYLE AND PETER WHEALE

1 Introduction

We have become used to spectacular scientific and technological 'successes' – putting a man on the moon, exploding a nuclear bomb, developing a new drug. It would be difficult to disentangle all the scientific theory behind both the hardware and the software in these programmes, but it would include gravitational theory, electromagnetism, quantum mechanics, relativity and molecular biology. We are inclined to believe that these great technical successes provide proofs of all the underlying theory. In a sense, each time we switch on a television set or use a calculator we appear to 'verify' a whole range of laws of physics, and in the end may be led to think of science as being as sure as reason itself.

On the other hand there are failures too – planes that crash, thalidomide children, nuclear power stations that generate only a small fraction of the output they were designed for. Disagreements occur frequently between technical experts, for example over the safety of a nuclear reactor, or the side-effects of a new drug. Passionate controversies between leading scientists sometimes surface in public. There seem, after all, to be uncertainties at the frontiers of scientific knowledge. Furthermore, we know that theories once firmly held and regarded as secure have later been 'falsified' and discarded.

This raises a series of philosophical and sociological questions. How 'true' is science? How valid and reliable are scientific 'facts'? How much faith should we have in what scientists say? How is scientific knowledge produced, and to what extent is it a social product? These questions, at first sight remote from the very concrete issues considered in Part II, do, as it turns out, have real practical significance. Section 2 treats science as a social activity, passing from the training of students to the role of the scientific community in the production of knowledge. Section 3 deals with some of the philosophical problems associated with the nature and growth of scientific knowledge. Finally, Section 4 discusses some of the difficulties involved in applying science to questions of practical concern to the wider community.

2　Science as a Social Activity

2.1　Training

We begin our consideration of science as a social activity by looking at the training of scientists. By scientists are meant people who are actively engaged in scientific research and who publish papers in the recognised scientific journals. Although there have been isolated men of genius in the past who were exceptions, a large majority of modern scientists has progressed from an undergraduate study of science to post-graduate work, usually for a doctorate, before finally at the post-doctorate level being recognised as fully-fledged scientists. Let us consider these stages in turn.

At the undergraduate level there are superficial similarities in the training of science and arts students – attendance at courses of lectures, assessment by examinations, and so on – but, in the science subjects more than most others, there is an emphasis on the learning by heart of masses of brute 'established facts' ('stenography plus memorisation', in Jevons' phrase (Jevons, 1973)), and on the manipulation through worked examples of key theoretical concepts. Lists of experiments have to be worked through in practical classes, with usually predetermined results to be achieved. In general there is a narrow rigidity with little opportunity for the individual expression that is encouraged in the best teaching of arts subjects – a range of views on D.H. Lawrence is permissible, but there is only one acceptable account of Kepler's laws (and that must be in modern dress).

For all the lip-service paid to the idea of the open, questioning mind, the successful students are the ones who, shutting out doubt and curiosity, practise assiduously on text-book examples, and concentrate on reproducing from books and lecture notes the sort of detailed information that scores high marks in examinations. Speculation on the historical origins, or on the validity of the concepts they use is unlikely to help them obtain a good degree. The science undergraduate (in spite of innovations in school science) is encouraged to see science as a body of hard, eternal facts, which are set out, always from the same single perspective, with very minor changes of form or order in the different text books currently available. History is suppressed or distorted, except in so far as it conforms to a picture of rational steady progress in the accumulation of facts and the formulation of ever better and more general theories to accommodate them.

Undergraduates are largely shielded from the problematic areas of science on the frontiers of research, though they are, of course, aware of their existence. It is not until the post-graduate stage that students come into contact directly with some of these problems, and try to make contributions to their solution.

What are the essential features of the training of PhD students? Perhaps the key to this question lies in their relationship to their supervisors, a relationship which has been compared to that of the apprentice to the master craftsman. Polanyi (1958) has drawn attention to the craft skills involved in science. Creativity and judgement of a high order at all levels are needed in good scientific work. These are tacit skills and are not any more likely to be acquired without outside help and practice than the ability to do elaborate carpentry or plumbing, or to play the violin well. These activities cannot be reduced to a formula to be applied by an individual in isolation. General rules for guidance and hints for avoiding pitfalls can of course be explicitly given, but much more than text-book knowledge is involved in the mastery of a craft and in its transmission. Sympathetic guidance from the expert is essential, with the disciple learning repeatedly from his mistakes, 'getting the feel' of what is permissible and possible.

At the same time as they are working directly on their research, typical young scientists are also absorbing the whole ethos of modern science. This is a process which had already started during their earlier studies, but now they will find themselves committed more and more to the system of values of the scientific community. They will come to justify their work in the words and phrases most scientists use. They will relate their own specialist areas to the total field of science, sharing the jokes of specialist colleagues against other scientific sub-groups, but at the same time looking outwards on the world from a consciously scientific viewpoint. In short, they will become familiar with (what we may call for lack of better terms) the mythology and ritual of the scientific community to which, finally, and if their work has been judged as satisfactory, they are admitted. We must, then, go on to consider the work of qualified scientists. As we shall see, they, no less than their students, are subject to social constraints.

2.2   The Production of Scientific Knowledge

At the outset we note one very elementary though very important point which is surprisingly often overlooked. Speech is social, and so therefore are words, which are evocative as well as descriptive. Science, like many other types of knowledge, is embodied in words and symbols, and transmitted through them. For this reason alone it must have a social dimension; it is not a mass of signals churned out by some omniscient, computerised automaton on a desert island, though some authors, to judge from their writings, seem almost to believe this.

Ravetz (1971) has given a very detailed and convincing account of the production of scientific knowledge. He sees the origins of this knowledge in scientists' attempts to solve problems they have posed themselves, using both new and suitably modified old concepts, methods and tools, both intellectual and practical. Applying all their skills, and making implicit value judgements at each step about the relevance and the validity of their work, they wrestle with their difficulties and eventually write papers, which in essence present arguments leading towards stated conclusions. (These papers are, of course, constructed formally, and by no means give chronological accounts of the authors' thought processes and activities – here again there is to be seen the tendency to present science as a collection of impersonal and objective absolute facts.)

It is at the point where a scientific paper is submitted to a journal that social control by the scientific community over the individual scientist is most clearly apparent. For the paper is assessed by referees who are respected members of the scientific establishment, and if it does not measure up to their standards of adequacy it is rejected. Since the publication of papers is of great importance to scientists' careers, it will be seen that there is a very strong pressure on them to present their work in such a way as to optimise its chances of publication.

Even after a paper is accepted, there is a great likelihood that it will be read (if at all) by a mere handful of people and then forgotten. Some few papers do, however, contain ideas that get taken up, developed, tested and applied elsewhere in the solution of other problems. Ultimately, though often in an almost unrecognisable form, an original idea from a paper may reappear in a text book where it is quoted as part of established scientific truth. But a long and tortuous road has been travelled by the concepts in Newton's *Principia* on their way to a modern treatise on mechanics. And it is a far cry from Ohm's original suggestion of 1826 (which contains no reference to current, potential difference or resistance – all concepts developed later) to a modern text-book statement of Ohm's Law.

Far from being simply the achievement of a few men of genius, scientific knowledge is the result of a long, complex and irregular social process. Just as there are stages when errors are eliminated, there are many other stages during this process when 'mistakes' creep in. Indeed, we may be sure that all mistakes and inconsistencies are never eradicated. No real scientific theory can be written down as a piece of formal logic to be regarded as proved for all time. Various writers such as Gödel in the 1920s have drawn attention to the limitations even of mathematics in this respect. The history of science and mathematics amply demonstrates that no scientific work is proof against inconsistency. One is continually reminded of attempts to stop water leaking from a colander which has more holes than one has fingers to block them. One or two leaks can be plugged here and there, but this only focuses attention on other holes previously unnoticed, where water now pours out.

Why is it that the social aspects of the production of scientific knowledge are so often overlooked or played down by scientists? At both school and undergraduate level, while the notion of teamwork may occasionally be alluded to, little real attention is paid to the slow evolution and development of such ideas as 'energy' or 'the ether', ideas full of obscurities which have been clarified or explained away in countless discussions over the years involving very many scientists and not just a few outstanding men and women. Even the recognised genius has to present his or her theories somewhere, and have them reformulated in different ways, debated, and finally (perhaps only with reluctance after a long period of hesitation) approved by distinguished scientific contemporaries. There is no higher court of appeal than these. In the last resort, physics, chemistry and biology are what the leading members of the scientific community declare them, at any moment in history, to be.

Perhaps this whole procedure is too reminiscent of the arguing and posturing of politicians of different parties with their views of what is, and is not, just and proper for society; and scientists feel that they are above this common herd. They are pursuing some absolute truth about whose value and possibility of attainment all are agreed. So we find them forgetting that, though constrained in some ways by the natural world, science is created completely by them, from the simplest concept they use to the most elaborate theory. Instead they would have us believe that they are 'asking questions of nature' rather than of one another; that they are merely lifting the curtain from some perfect picture that has been in existence for all time, rather than (as all groups of thinking men have always been doing) painting for themselves a somewhat confused canvas, which takes on quite different appearances in different cultures and in different ages.

Sociologists of science have examined the ways in which scientists actually perform their work, and how far scientific knowledge can be considered a product of social life. The American sociologist Robert Merton has described the 'ethos of science' by using a framework in which the scientific community is characterised by four norms which he believes form a code of conduct for scientists (Merton, 1957). These norms, he suggests, act as rules which are seen to govern the behaviour of scientists once they are adequately socialised and thus enabled to perform their roles. These four norms are: universalism, communalism, disinterestedness and organised scepticism (where universalism refers to truth claims; communalism refers to the common heritage of the substantive findings of science; disinterestedness describes the impartial nature of scientists' research activities; and organised scepticism implies the suspension of the scientists' judgement until 'the facts are at hand', and a detached scrutiny of beliefs, in the face of empirical and logical criteria, has been extensively performed).

Once socialised, scientists are considered never to have difficulty in agreeing about the status of relevant knowledge. Because Merton views scientists as role-playing (that is, their conduct in their positions at work is governed by a set of accepted rules), he believes they are able to identify 'normal' behaviour and 'deviant' behaviour. This form of analysis of scientists' activities implies the closed nature of scientific culture, which has led Merton to emphasise an 'internal versus external' mode of analysis.

In practice, scientists frequently deviate from these supposed norms, indicating that, in fact, they are not firmly institutionalised. Mitroff (1974), for example, argues from evidence based on his studies that scientists often have great emotional commitment to their hypotheses and theories, that secrecy concerning research findings is common, and that their judgements of knowledge claims are often based on personality and prestige rather than on objective assessment. (See also Sklair, 1973.)

Merton's analysis of the scientific community also inadequately treats change and innovation. If scientists operate in groups with closed cultures consisting of sets of norms, how is radical change of thought and action ever brought about? Merton attempts to answer this question by invoking the 'ethos of science' as itself fostering an appropriate climate for innovation, but this is not a convincing explanation in his scheme of sociological analysis. (See Law and French, 1974.)

Kuhn (1970) rejects Merton's analysis of the production of knowledge, and bases his own work on studies in the history of science. He also rejects (see page 36) the characterisation of the open-minded and uncommitted scientist testing his hypotheses, and instead suggests that scientists, by virtue of their training and apprenticeship in an established scientific discipline, become committed to particular ways of viewing their subject-matter and of arriving at explanations. Scientists, Kuhn argues, are socialised into particular academic cultures and develop their own 'scientific communities', working mostly in consensus groups, each basing its work on a shared paradigm. Change in scientific thinking, according to Kuhn, occurs when sufficient anomalies between fact and theory have accumulated to cause a paradigm 'crisis', which eventually results in a new paradigm.

Kuhn's mode of analysis embraces the specific activities, theories and concepts of the scientist and has been taken up enthusiastically by a number of sociologists of science. Crane (1981), for example, in a study of physicists working in the field of theoretical high energy physics, concludes:

What holds this large and complex field together is the commitment of its members to a set of remarkably durable fundamental principles . . . this gives them a common perspective and a common language, regardless of how diverse their specific interests may seem.

Mulkay (1972, 1979) has suggested a more convincing explanation of the innovative process than Merton by arguing the case that scientists are predisposed to innovate when they occupy different roles in the scientific community ('role hybridisation'), when they occupy peripheral status in their scientific groups ('marginality'), or when there is fierce competition between scientists for publication of research work. His studies suggest that it is the clash of beliefs and theories of knowledge, and the transfer of these norms from one scientific group to another, that account for the process of innovation. The implication of this is that innovation often arises through 'scientific invasion' – the transfer of normative frameworks from one field of science to another.

Kuhnian case studies, by suggesting that the products of scientific activity are shaped by social factors, promote scepticism concerning the impartial status of scientific knowledge. An implication of this is that scientific knowledge may be relative, not absolute, and thus to some extent culture-specific: relativism is at present a hotly debated question in the philosophy of science.

Studies by Collins (1974, 1975) of groups of scientists researching on lasers and gravitational waves provide evidence of the social nature of the production of scientific knowledge. He argues that there is often no common assessment of experimental procedures or results by group participants, and that scientists 'negotiate' judgements concerning the merit of the knowledge claims involved.

Control by professional groups on the legitimisation of knowledge and the dissemination of relevant information is also important in determining what is to be considered 'valid' knowledge. Thus Frankel's (1976) study of developments in optics in the nineteenth century argues that the rapid rise to dominance of the wave theory of light in France was greatly enhanced by the emerging dominance of an anti-Laplace faction (Laplace and his supporters adhered to the corpuscular theory of light) in the French scientific community. The members of this newly emerging dominant faction were able to exploit their powerful positions in teaching, research and publishing to convert the next generation of students to their view.

Such studies as those cited above have indicated that questions of the acquisition and use of information by scientists are not separable from questions of the production of knowledge, and have emphasised the study of what questions produce 'meaningful' responses from scientists. Mulkay (1979) provides an excellent review of a number of these 'interpretative' sociological case studies, and also points out that many sociologists and philosophers of science have converged on a conception of science as an interpretative enterprise where what is to be considered as scientifically-valid knowledge is 'negotiated' by the scientific community, and is not simply empirically discovered. Scientific knowledge, according to this view, is conditioned by the socio-political and economic context in which it is produced. Thus, the personality and reputation of scientists, and the status of the professional bodies who determine and disseminate scientific knowledge, are seen as important factors which must be taken into account if a more comprehensive understanding of our body of scientific knowledge is to be achieved.

3 The Nature and Growth of Scientific Knowledge

We start with a view of science which most philosophers of science consider to be totally discredited. But – to judge from frequent statements – it is still held by surprisingly large numbers of working scientists, as well as by others who know nothing of science. It is often what people have in mind when they talk of the 'scientific method'.

This account of science, in its most naive form, describes experimenters making observations, collecting facts using both qualitative and quantitative techniques, then making hypotheses, testing predictions made on the basis of these hypotheses, and rejecting them if the predictions are not verified. If they are verified, gradually the numbers of valid hypotheses increase and cohere into more general hypotheses and theories. This picture of scientific knowledge makes it seem like a great heap of sifted sand, to which geniuses contribute huge bucketfuls and minor figures mere spoonfuls as their life's work. As the pile grows it sets and becomes more solid.

There are many difficulties about this notion. First, making observations and collecting facts are problematical activities, and they make no sense unless they are carried out in a particular context. To observe is first of all to select, to apply consciously or unconsciously certain criteria of relevance about what it is worth while to observe. What we observe is also heavily influenced by what we expect – in other words, observations are theory-laden, and can only be expressed in terms that are rooted in theory. Take the simple statement: 'The current was observed to be 2.75 amperes.' In fact, someone may have seen a pointer moving on a scale to indicate 2.75; the terms 'pointer' and 'scale' would not even be used by a person unfamiliar with our culture and its electrical equipment. The original statement takes for granted the whole theory of electromagnetism underlying the operation of the ammeter used, as well as the concept of current and its units, amperes. 'Scientific facts', like 'observations', are also bound up with theories and preconceptions, and must be considered in context. We argued in section 2 that scientific facts do not spring suddenly to light, but emerge, rather, at the end of a long social process. Facts too are selective. We shall also argue later that because of the focus of science on certain parts and aspects of human experience at the expense of others, it can never provide the whole truth about a situation.

A second difficulty about the naive view of science relates to the question of drawing conclusions from observations – the so-called process of induction. We note that, although repeated observations and measurements giving consistent results may point strongly to a certain conclusion, this is not the same as proving that this conclusion is logically necessary. No number of cases of A being B can ever establish that all As are B. The paradox about scientific knowledge is that, although it is provisional, not logically secure, and subject to question and revision, nevertheless it is felt to be surer than other forms of knowledge. We must consider why this should be so.

Thirdly, it is not accurate to say that the growth of science is a simple accumulation of knowledge, analagous to an increasing heap of sand. New developments often involve some sort of fundamental reorganisation of ideas, with new concepts and insights emerging, that provide new emphases, and lead to the banishment to the periphery of questions formerly considered central. One might think of the conceptual changes associated with the shift from the classical physics of the late nineteenth century to the quantum mechanics and relativity theory elaborated early this century. Certain ideas like 'the ether' vanished, others like 'mass' and 'energy' were redefined, and new concepts such as 'quantum' and 'transition probability' appeared. It was not just a matter of adding new pieces to the old framework, but of replacing the framework itself.

Popper, whose work first became widely read in the 1960s, attacked the problems associated with induction by switching the emphasis from verification to falsification. What the good scientist does, according to him, is to put forward the boldest hypotheses possible, and then try to falsify them. Science is a matter of 'conjectures and refutations' (to use the title of one of his books) (Popper, 1964), only if a theory stands up against powerful attacks on it, can we have some confidence in it. It is a matter of great importance, of course, that theories are, in principle, falsifiable. Indeed, what distinguishes genuine scientific theory in physics, for example, from Freudian or Marxist theory (which by Popper's criterion is unscientific) is its 'falsifiability'. Popper's scientific methodology is an example of the 'hypothetico-deductive' method, and he contends that a scientific theory can only be accorded provisional acceptance, that is, a theory holds until it is falsified. Thus *falsification* is held to be the observational and experimental procedure of science.

Things, however, are not quite so simple as is suggested by this 'naive falsificationist' position, as it is called by its critics. The same sorts of techniques of protective adjustment and auxiliary hypotheses, which, according to Popper, make the works of Freud and Marx, for example, resilient or unassailable, but invalid as science, can also be found in what he regards as science. All scientific theories when first put forward fail to account for all the experimental evidence; contradicted on several sides, they are born 'refuted', but luckily they are not rejected out of hand. As an example, we may take Bohr's early model of the atom, with electrons moving in stable, circular orbits around the nucleus. This was considered impossible according to classical physics, because the electrons would radiate, lose energy and spiral inwards. Nevertheless Bohr stuck to his model, and it proved very useful in the development of quantum theory.

Lakatos, to counter these criticisms, developed Popper's ideas by maintaining that proponents of a scientific theory needed to be tenacious. It was a question of 'competing research programmes' supporting rival theories. These programmes, devised to tackle and iron out difficulties, could be 'degenerating', leading nowhere, or 'progressive', leading to consolidation and clarification of the theory (Lakatos and Musgrave, 1970).

Popper and Lakatos were at pains to defend the rationality of the enterprise of science against a variety of critics of this rationality. One such critic was Kuhn, the American historian of science, author of perhaps the most influential and widely-read book in the field of the philosophy and sociology of science in the post-war period – *The Structure of Scientific Revolutions*, first published in 1962 and in an enlarged edition in 1970. Kuhn distinguishes between *'normal science'* and *'scientific revolutions'*. The first term, which refers to what most scientists are doing most of the time, is used to describe the activity of researchers working at *'puzzle-solving'* within a well-established tradition or *'paradigm'*. A paradigm, in one of the senses used by Kuhn, is a group's 'constellation of beliefs, values, techniques, and so on', in relation to which and in terms of which, research problems are defined, investigated and solved. A physicist of the last century, looking at the properties of a semi-conductor, say, would make uncritical use of the knowledge and ideas then available to him from his training, experience, the advice of colleagues and the specialised literature of the day, all this amounting to a paradigm very different from the present one. Research work does not continually seek to undermine established theory; it is puzzle-solving in accordance with given rules. This, says Kuhn, is normal science. If a scientist obtains anomalous results, it is not the basic theory which is called into question, but rather the abilities of the researcher.

Historical studies show, however, that in all scientific fields, after a stable period, unanswered queries accumulate, anomalies build up which cannot be attributed to incompetent research. There is a period of great activity and instability when a new theory is proposed which requires a reorientation, a re-ordering of experience, like the *gestalt* switch required to see in the well-known drawing, at one moment a set of cubes viewed from above, at the next the set viewed from below. This process of suddenly 'seeing the light' has something in common with a religious conversion. Kuhn quotes Planck's remark to the effect that opponents of a new theory (such as quantum theory in the early 1900s), do not often become converts, but rather die off gradually over a period. Admissions of past error by leading scientists and whole-hearted immediate acceptance of the new theory are not common. In the scientific, as in other communities, there is resistance to change.

These then are the processes involved in scientific revolutions, of which the biggest and best examples are the overthrow of Ptolemy's geocentric – by Copernicus's heliocentric – system in the sixteenth century, of phlogiston theory by Lavoisier's new chemistry, of previous theories of origin by those of Darwin and Wallace, of Newtonian by relativistic mechanics, and of classical physics by quantum theory. The latter two relatively recent revolutions are particularly spectacular examples of great discontinuities in the development of physics, which even before 1900 was regarded as very securely established.

As we have seen, Popper's 'demarcation criterion' – his rule for distinguishing genuine science from other forms of knowledge – hinges on the concept of falsifiability. For Kuhn, it has to do with the presence or absence of a paradigm in the area in question. In the field of electricity in the eighteenth century, for instance, there was general disagreement about fundamentals; there were no theories or concepts or text books accepted by all practitioners and schools of thought as a base for further work – in short, no paradigm existed. By 1850, however, the situation had changed. New experimental and theoretical work had led to sifting, modification and reconciliation of the earlier diverse views so that a basis for agreement was reached; electricity theory had become 'scientific'.

The idea of the supreme rationality of science has been even more deeply questioned by Feyerabend (1975), a self-styled 'methodological anarchist' with the slogan 'anything goes'. His advice to scientists is to champion not the most apparently 'reasonable', but the most outlandish theories available that fly in the face of the facts! This, argues Feyerabend, is what was done by Galileo, for example, in the early seventeenth century, when he laid the foundations of mechanics, and this is what is necessary today to revitalise modern science, which has become a tedious and conformist subject.

In works on the philosophy of science there is often confusion between *description* and *prescription*. There is conflict between writers who purport to base their accounts of science on what scientists actually do, and those who say what ought to be done, and discuss some idealised or trivialised form of science, which has been tightened up logically. Some philosophers focus their attention on a static picture of science, treating it as a fixed body of knowledge; others emphasise its capacity for reorganisation and development. It is noteworthy, too, that there is a relative neglect, on the one hand of the biological sciences, and on the other of technology.

Most of the main figures involved in the Kuhn-Popper debate have drawn their examples largely from the pure physical sciences, and especially from physics.

The philosophy of science has interesting implications for theorists of disciplines, such as economics and sociology. Ever since the days of Comte in the nineteenth century there have been those who argue that the social sciences should, as far as possible, imitate the approach and methods of physics or the models of systems analysis; and others who strongly disagree with this positivist approach. Popper's ideas on science have led him to elaborate a theory of society which advocates piecemeal social engineering and evolutionary social change. Radical thinkers, taking the opposite line of argument, have seized on certain aspects of Kuhn's work, and have insisted on the importance of revolutionary changes in society as in science.

To conclude this section we return to the questions posed in the introduction to this chapter: How true is science? How valid and reliable are scientific facts? We have seen that science cannot be regarded as absolute, incontrovertible truth; it is best regarded as provisional knowledge, and, if the past is any guide, even the theories which we believe at present are established on the surest foundations may well be overturned in the future by some new and totally unexpected developments. Technologies have been established in the past using flawed or mistaken scientific doctrines, so even a successful technology gives no guarantee that the underlying science is 'correct', though we may well have great confidence in some apparently well-tested theories.

We shall remember too that many people are hostile to science or reject it. We discuss their views in more detail in Chapter 12, section 3, and here simply state that these critics feel that science, in its emphasis on mathematical analysis and objectivity, necessarily distorts reality by deliberately excluding emotions and subjective, individual reactions to events – all, they say, that is most valuable and important in human experience: thus science presents a de-humanised view of the world.

We attempt to summarise these opposing views on science by using an analogy between scientific knowledge and a map. Science can only 'work' if there is some degree of similarity between at least some features of the mathematical-analytic-scientific 'map' of reality and the corresponding features of reality itself. The fact that science 'works' – that aeroplanes fly, that television sets can be tuned to receive pictures transmitted from the moon – does not imply, of course, that the whole scientific map is accurate. To pursue the metaphor – vast tracts of unfamiliar land rarely visited may be wrongly marked without, for a time, causing too much inconvenience. On the other hand, even those sections of the map about which most confidence is felt, may either be lacking in detail or overloaded with symbols that have no counterpart in reality. Sometimes a crude and inaccurate map is better than no map at all; at other times, it is worse, it will lead one even farther astray.

Just as the most useful map for the traveller is one that marks in roads and railways, leaving out other features, and is quite different from a map serving the purposes of, say, the geologist, so science concentrates on certain aspects of reality, of the world, and necessarily neglects others, equally if not more important for our lives and our relationships with our fellows. 'Scientific truth' is only partial truth, and is necessarily contingent, ultimately 'unprovable'. It is a view of the world from one angle: a full account of phenomena must include very much more than 'scientific explanation'.

4 Applied Science and Technology

The demarcation between pure and applied science is very vague. It is often difficult to say to which area a particular piece of research belongs. Work may start as 'pure' and end up as 'applied', and vice versa. A training in pure science can equip a man to become a good technologist. In debates about the relative funds which should be assigned to pure and applied research confusion arises because of the differing emphases placed on these terms by different authorities. Nevertheless, there are some clear distinctions, and it is instructive to look at a few of the features of applied science and technology which set them apart from pure science. For this purpose we go back to the discussion of problem solving and again use ideas suggested by Ravetz (1971). It should be noted that many of the general statements made below are illustrated by particular examples in Part II.

Ravetz distinguishes two different types of problems which are encountered when scientific knowledge is applied – *technical* and *practical*. By a technical problem he means one in which an externally assigned function has to be performed (e.g. the manufacture of a rocket to land a man on the moon, or indeed the manufacture of tools in general). Practical problems are usually the most difficult and most complex; the goal in solving them is to achieve some human purpose such as, perhaps, providing welfare of some sort for a particular section of society. Thus the provision of housing, education, energy, employment are all practical problems. Large-scale practical problems often spawn subsidiary technical ones. We might think, for example, of the complex practical and technical issues associated with one large-scale practical problem – the future provision of energy for the UK. Here only a few general remarks are made.

In many ways technical problems pose a greater challenge than purely scientific ones. For one thing there may be a certain urgency about their solution, and they cannot simply be left aside, as a piece of pure research might be if the results are not felt to be promising or interesting. The pure scientist has more freedom also in the sense that he can usually narrow down the number of variables he has to consider simultaneously, while this may be impossible when a technical problem is under consideration and certain specifications have to be satisfied. The technologist will have to bear in mind factors such as cost, marketing and servicing as well as the technical specifications. He will have to choose between conflicting requirements; compromises in design will be necessary to achieve a useful synthesis; an immediate, imperfect though workable solution will often be preferred to the prospect of a much better one in the distant future. It follows that great flair, judgement and flexibility of thought as well as good scientific understanding are needed for the successful solution of technical problems.

The criteria by which success is judged are also not the same as for pure science, since the public may find it difficult to assess the future relevance or importance of a particular piece of research. In principle, if the machine designed or the software produced as the solution of a technical problem, carries out the functions listed in the specifications laid down at the beginning, then success has been achieved. Often this will mean that goods are produced which are commercially profitable and will compete advantageously with similar goods from rival companies. In certain cases, however, especially when only one company is involved in a large-scale project, commercial restraints do not apply, as, for example, in the design and manufacture of weapons systems or new types of aeroplane such as Concorde. Here questions of national defence or prestige come into play and cloud the issues. Experts called in to assess the efficiency of the end-products may well take up attitudes determined by other than purely scientific considerations. Pressures may be brought to bear, values become distorted, and, with enormous sums of money at stake, bribery or other forms of corruption may occur. 'Runaway' technology, then (as the solution of technical problems has aptly been called when costs escalate far beyond original estimates), is a common danger when commercial prospects are irrelevant or set aside.

If technical problems generally have extra dimensions that purely scientific ones do not possess, practical problems in their turn involve a whole complex of social and political considerations which make even difficult technical problems seem simple by comparison. Even when there is a consensus about the ultimate purpose to be achieved (and this is rare), there will always be disagreement about the means to achieve it. Various groups in society will see their interests threatened in various ways, or will be eager to exploit situations arising, in order to extend their influence or extort financial profits. The problem will be defined and stated in different terms, corresponding to differing philosophical and political beliefs, and each different statement will imply a different line of action to be initiated to bring about a solution. Some people will attempt to reduce it to a series of purely technical problems, but in fact no practical problem can be reduced in this way – there is always a question of interests involved. Even when a decision on how to proceed has been taken, as time goes on there will be more or less subtle shifts away from the original ultimate goal, often brought about by bureaucracies which wish to impose their own form upon the solutions proposed. Very serious technical difficulties may arise, perhaps because of unexpected side-effects, or political changes may occur, so that the perceived problem and the proposed method of solution may end up almost totally transformed. In any case, due to the very nature of political debate, there will never be complete agreement about whether or not the original problem has been solved or to what extent the goals have successfully been achieved. (Questions relating to the control of technology are taken up in more detail in Part III of the book.)

References

Collins, H.M. (1974), 'The *TEA* set: tacit knowledge and scientific networks', *Science Studies* vol. 4, 165–85.

Collins, H.M. (1975), 'The seven sexes: a study in the sociology of a phenomenon, or the replication of experiment in physics', *Sociology,* Vol. 9, 205–24.

Crane, D. (1981), 'An exploratory study of Kuhnian paradigms in theoretical high energy physics', *Social Studies of Science,* vol. 23, 23–54.

Feyerabend, P. (1975), *Against Method* (London: New Left Books).

Frankel, E. (1976), 'Corpuscular optics and the wave theory of light: the science and politics of a revolution in physics', *Social Studies of Science,* vol. 6, 141-84.

Jevons, F.R. (1973), *Science Observed* (London: Allen & Unwin).

Kuhn, T.S. (1970), *The Structure of Scientific Revolutions* (Chicago: University of Chicago Press).

Lakatos, I. and Musgrave, A. (eds.) (1970), *Criticism and the Growth of Knowledge* (London: Cambridge University Press).

Law, J. and French, D. (1974), 'Normative and interpretive sociologies of science', *Sociological Review,* Vol. 22, 581–95.

Merton, R. (1957), *Social Theory and Social Structure* (Glencoe, Ill.: Free Press).

Mitroff, I. I. (1974), *The Subjective Side of Science* (Amsterdam: Elsevier).

Mulkay, M.J. (1972), *The Social Process of Innovation* (London: Macmillan).

Mulkay, M.J. (1979), *Science and the Sociology of Knowledge* (London: Allen & Unwin).

Polanyi, M. (1958), *Personal Knowledge* (Chicago: University of Chicago Press).

Popper, K.R. (1969), *Conjectures and Refutations* (London: Routledge & Kegan Paul).

Ravetz, J.R. (1971), *Scientific Knowledge and its Social Problems* (Oxford: Oxford University Press).

Sklair, L. (1973), *Organized Knowledge* (St Albans: Paladin).

4 Politics of Science and Technology

PETER WHEALE

1 Introduction

The social context in which scientific activities are pursued has a political dimension: it is this dimension which we shall explore in relationship to science and technology in this chapter.

Whenever people must adopt a group policy on any particular matter, and have differences of opinion on exactly what that policy should be and how it should be carried out, there arises a clash of interests. This clash of interests is the essence of politics. Conflict implies social power, and social power can take many forms, from coercion and manipulation, to bargaining and persuasion. When social power is formalised within institutions such as the Church or the Law, then the possibility arises that conflict will not be apparent. The predisposition of a social system can allow some interests to dominate over others, creating 'false consensus' in which resistance is subsumed in the socio-economic structure and functioning of the social system. (See Partridge, 1963.)

Technology is often associated with social change and theories of social change are considered in Section 2. Section 3 examines the political effects which the scientific world-view is claimed to have in industrial society. The developments in science policy are reviewed in Section 4, where we examine the idea that there has been a 'politicalisation' of science which has resulted in a schism in the relationship between science and society. Section 5 critically considers the notion that there has also been a scientisation of politics. The chapter concludes with a brief discussion of the potential effectiveness of grass-roots political action directed at influencing national science policy, and a review of the case against science and technology as oppressive forces in society.

2 The Politics of Social and Technical Change

Many thinkers have associated science and technology with the process of social change. It is often suggested that the Industrial Revolution occurred as a direct consequence of our use of innovations such as steam power, improved methods of iron and steel production, textile machinery, machine tools and so on. The US economist Galbraith (1969), for example, argues a thesis close to this 'technological determinist' view of social change. He considers society as responding to and being moulded by the 'technological imperative'. In his view, the industrialised nations' use of high technology produces a society of large bureaucratic organisations requiring vast amounts of capital to run them. Galbraith suggests that industrialised society is characterised by its stage of technological development, and that there is a broad convergence between all industrial societies despite their differing political and economic systems. This analysis has the weakness, however, that it gives no explanation of why particular technological innovations occurred at particular times, or why many societies failed to use the technical knowledge which they possessed.

Karl Marx (1818 – 83) developed the historical materialist theory of social change. The idea here is that society is essentially dynamic rather than static, it emphasises conflict rather than consensus, and considers that changes in the social structure occur as a result of technological changes and their economic consequences. Marx believed that, just as Darwin had developed a scientific theory of natural evolution, so his theory of change and revolution identified 'laws' of social evolution. The materialist conception of history has led Marxist historians to consider the correspondence between particular social and economic structures and the type of technologies deployed by society. Thus they saw clear links between the factory-based technologies of the Industrial Revolution and the capitalist economy. As we saw in Chapter 2, some Marxist historians have argued that not only technology, but also science, both in its directions of development and in its content, is closely tied in to social and economic structures. Much interest, for example, was generated in Europe when the Russian, Hessen (1931), argued at a conference that even Newton's *Principia* could only be fully understood when seen in the light of the economic and technological needs of English society in the late seventeenth century.

Marx (1859) considered capitalist society to consist of three main strata: the upper class (or ruling class), the middle class (or bourgeoisie), and the working class (or proleteriat). When a particular stage of history is mature the forces of production (technology) are used by the ruling class to sustain their power and control over the social order. However, although technology is not therefore politically 'neutral', as it is renewed and developed the ruling class is unable to manipulate the new technology to its advantage, at least in the long term. Continued control by the existing ruling class is undermined and a new group of persons associated with the emerging technology rises to displace the presently existing ruling class (whose power has been dependent on its control and ownership of the now outmoded technology). Marx argued that just as the hand mill had engendered a feudal society, and the steam mill an industrial capitalist society, so a new scientific and technological age would eventually produce a proletarian revolution, and a communist society.

Burnham (1941) took up a modified version of the Marxist theory of social change. He agreed with Marx that control of production provides the controllers with political power and social prestige as well as wealth, and that the form of this control is determined by technological and economic forces. However, Burnham's thesis was that the productive process has come into the hands of a managerial class, a skilled technical élite, whose position is dependent on technical expertise. He argued that the capitalist class would ultimately be displaced, not by the proletariat as Marx believed, but by an indispensable technical and managerial élite. Burnham's view is similar to that of the French utopian thinker, Saint-Simon (1760-1825), who coined the term 'industrialism', and who considered that society would eventually become a scientific association, in which wealth would be created by mechanised production, and which would be governed by a meritocracy of technologists and scientists.

Some critics of the Marxist 'conflict' theory of social change take a functionalist approach to understanding social developments. Functionalism has been mostly applied by sociologists and social anthropologists, and uses a conceptual framework which relies on the idea that the social structure is normally essentially cohesive and stable, and consists of integrated parts. Furthermore, it considers society to be purposeful (or teleological) and tends to argue that society has 'needs'. The 'ideal' state or condition of society is perceived by functionalists as one of 'equilibrium' or 'homeostasis'; and conflict or 'disequilibrium' causing change tends to be seen as 'abnormal', 'dysfunctional', or 'pathological'. Implicitly, at least, social stability is held to be beneficial for society, whilst instability (conflict, disruption, political revolution), is held to be harmful.

Durkheim (1858–1917), the early French pioneer of functionalist sociology, was concerned with the study of cohesion and the maintenance of order in society. He believed it was the function of moral and legal 'rules' to constrain disruptive behaviour. Durkheim (1898) exemplified the integrated character of society by describing the social interdependence brought about by the increasing specialisation of labour required by industrial society. A more recent writer to adopt a functionalist approach to the analysis of industrialised society is Smelser (1959). Smelser applied the concepts of 'adaptation', 'interdependence', and 'equilibrium' to describe the development of the Lancashire cotton industry in England during the period 1770–1840. He views this period as dynamic for this industry, a new equilibrium for the industry only being achieved after 1840. He argues that the new equilibrium in this industry was brought about by its structure adjusting to changed conditions; that is to say, an economy consists of parts, structurally differentiated, all of which are adjusting to each other under changing conditions to produce a new equilibrium and stability as a whole.

The influence of biological ideas is implicit in much of the thinking applied in the functionalist approach to understanding social change. Society is often seen as analogous to a living organism evolving through time, and adapting to changed conditions in order to survive. The equivalents to biological organs are the institutions of society such as the family, the economy, and political bodies, all of which have 'functions' and are considered to evolve from the simple to the complex over time. Functionalists tend also to be committed to an 'empiricist' view of the world, that is, they believe that it is possible to build up scientific theories logically in coherent frameworks, and to relate them to 'social facts' in an unproblematic way.

The *a priori* assumptions held by 'technological determinists', 'conflict' theorists and 'functionalists' tend to predetermine not only their descriptions and explanations of social change, but also their *prescriptions* for society. Marxist theorists tend to favour 'revolutionary' change, whilst functionalists emphasise social order, incremental change and stability. It may be argued, therefore, that these approaches, whether conflict or functionalist, as applied to social analysis, have the quality of being normative (value-laden) because their *a priori* assumptions predispose the respective researchers to draw out particular explanations, predictions and prescriptions for society.

3   Ideology and Industrial Society

Political determinism is an approach to the analysis of society which considers that political forces are the main factor in determining the form and even sometimes the content of scientific and technological developments. When the Nazis came to power in Germany they instigated a purge of scientists. Many Communist and Jewish scientists were deemed unfit to serve the Third Reich and were dismissed. Einstein's revolutionary work on relativity was denounced as 'Jewish physics', and in universities and scientific establishments those sympathetic to his work were removed. Cruel racist experiments were also carried out in which it was hoped to demonstrate the superiority of the Aryan race over other 'inferior breeds'. (See chapter 7 for further discussion of racism in science.)

The Lysenkoist movement of the 1930s to 1960s in the Soviet Union is a further notorious example of a failed attempt to direct scientific developments, in this instance the application of genetics. After the Revolution of 1917 the Soviet Union adopted a form of Marxist ideology, which includes the view that human nature is in constant flux, and is historically determined by the particular society in which it exists. The new state wanted science to progress in such ways as were compatible with this Marxist view. Mendelian genetics (discussed in chapter 7) was not thought by the regime to be so compatible; whilst the Lamarckian theory of inheritance was thought to be so. Lamarck (1744–1829) had stated that characteristics acquired during an organism's life-time, in response to its environment, may be transmitted to its offspring, and Lysenko's research work, which attempted to utilise Lamarckian ideas, was received with enthusiasm by the Soviet authorities. The Politburo, with Stalin's support, assisted Lysenko to gain control of state research facilities from more orthodox (Mendelian) geneticists. Lysenko's research aimed at the directed transformation of varieties of plants by means of environmental manipulation and grafting. This proved to be a scientific failure and had disastrous consequences for Soviet biology as a whole, and many scientists personally (see Lewontin and Levins, 1976).

The racist experiments in Nazi Germany and Lysenkoism in the Soviet Union failed to impose particular political ideologies on scientific developments because these ideologies were incommensurable with the natural world – that is, they led to predictions which were clearly not fulfilled. As we have hinted in Chapter 3, scientific theories are always underdetermined by observation-based 'facts' – in other words, at any time on the frontiers of science, there are usually several theories which, together with auxiliary hypotheses, will account for the known 'facts'. Some of these theories may be favoured more than others by a particular ideology, and thus it is possible for social factors to influence the form and even the content of science.

Radical thinkers have begun to give detailed consideration to the possible political functions of technologies in society. This view contrasts with the conventional belief that technology is a politically-neutral force favouring no particular class or groups in society. It is argued (see, for example, Marcuse, 1974) that this conventional view is derived from the philosophy of 'scientism'. Scientism arises when it is believed that the scientific approach is objective, and therefore the only worthwhile approach to any problem. Scientism, it is claimed, promotes a passive acceptance of techniques and technologies which are perceived to be scientifically based and thus apolitical. Dickson (1974), for example, claims that technological neutrality is a myth, an 'ideological disguise', used to legitimate the continuing power of the dominant class, and that the technological determinist thesis in history is also an ideology. He discusses the setting-up of factories for weavers in Britain in the 1760s, and argues that this change, from what had previously been a 'cottage industry', was instituted by the merchants better to control the weavers' work. Output was increased by forcing the weavers to toil longer hours at greater speeds, and all technical innovations were regulated in order to ensure that they were applied solely for capital accumulation. The form of this technology, Dickson believes, derives from what he terms the 'ideology of industrialisation', which he considers is a political ideology of control, and enables employers to use, for example, assembly-line production techniques as a means of maintaining hierarchical relationships; television as a means of centrally controlling public information and entertainment; and, 'built-in obsolescence' as a means of manipulating consumer choice in order to stimulate the flow of market commodities. Existing communist societies are as guilty as the West, it is argued, because their present technology is permeated by the same exploitative ideology.

His argument is that no amount of *alternative* technology will help humanity, unless there is first a political and social revolution. This view may appear to conform to the Marxist doctrine, but on close inspection it diverges in a fundamental way from Marx's theory of revolutionary change. Marx believed technology was the agent of change which ultimately was not susceptible to manipulation by the dominant group (ruling class), and when the forces of production (technology) were conducive to proletarian revolution, then it was for the 'proletariat' to take control. The historical process would eventually lead to the displacing of the capitalist class by the proletariat and a communist society would evolve. Contemporary radical thinkers such as Dickson adopt an élitist theory of social change by maintaining that the ruling class can ensure that technology develops in such a way as to lead to their continued dominance in society. If a particular class or group in society *is* capable of manipulating technological and social developments, then how much more difficult must it be for any oppressed class to envisage, let alone perpetrate, a political or social liberation!

4   Science Policy

Science and the scientific world-view had become important forces by the seventeenth century. True to the spirit of the Renaissance, science, because of its specialised nature, was élitist, being practised by individuals who were not accountable to the general public. The relationship between science and society was almost from the outset divisive. Whilst the public was generally ill-informed of the work of scientists, scientists themselves created scientific societies which facilitated national and international communication. The Italian Accademia dei Lincei (1603), the British Royal Society (1662), the

French Academy of Sciences (1666), and the Berlin Academy of Sciences (1700) were amongst the first. By 1863, when the National Academy of Science was established to advise the American Federal government, several hundred scientific societies existed. A 'gentlemanly amateurism' characterised British science, at least before the First World War, which contrasted with the professionalism of the more state-linked academies of many other industrial countries. In Britain (in contrast to France, for example) no salaries were paid to members of the Royal Society, and thus until the nineteenth century a scientific career was only possible to those with private incomes. (See H. and S. Rose, 1969.)

Scientific activity was not directly associated with the class conflict that developed during the nineteenth century in the industrialised nations. However, the emergence of nation-states and the rapid industrialisation of Northern Europe and the US, made imperative the politicalisation of science, and the adaptation of the scientific community to the increasing plurality and social differentiation that accompanied industrialisation. The development of nation-states required governments to preoccupy themselves with their sovereign integrity and the technological means by which it could be maintained. Scientists had become, by the early twentieth century, a national resource.

Armytage (1965) points out that the British government, until just before the First World War, failed to appreciate the extent of the scientific and technological underpinning of society. War revealed how crucial the alliance between state and scientist had become. The British were soon at pains to create government-funded research councils to form the basis of scientific organisation, and to ensure that adequate resources were made available for scientists to pursue their researches in military, medical, agricultural and industrial fields. Science advisory committees were established in 1914, and only then did Britain have the beginnings of an integrated science policy.

Science and technology have greatly increased the problems of government. Because their effects permeate the whole social system they are increasingly seen as defining or complicating the substance of public matters. (See Williams, 1971.) It has been suggested that there is a tendency towards the 'scientisation of politics'. Habermas (1971) argues that the politician is becoming the 'mere agent of a scientific intelligentsia', and it must be acknow-

ledged that technical decisions are made every day, at a bureaucratic level, by experts who are not held democratically accountable. Lindberg and Scheingold (1970), when discussing the early
Common Market wrote: 'it would not be an exaggeration to characterise the entire community as essentially bureaucratic and technocratic.'

Mobilisation of scientists and engineers in the Second World War
highlighted the need for a central advisory body to formulate
government scientific policy, and after the war acute problems
arose over the issue of the allocation of resources for research and
development. There were conflicting demands from different
laboratories and research groups, and competing fields, such as
high energy physics and molecular biology (Gummett, 1980).
Science policy in Britain, for example, was further controlled by the
setting-up of an Advisory Council on Scientific Policy at the beginning of 1947. It was decided that the Department of State, with
executive responsibilities, should be responsible for identifying
problems requiring research, stating their order of priority, and
deciding where to carry out the work, and how to apply the results:
the research councils were to continue to initiate background
research which would be free from administrative control. (These
are now the Science and Engineering, the Medical, the Agricultural, the National Environment and the Social Science Research
Councils, known usually by the acronyms: SERC, MRC, ARC,
NERC and SSRC.) In the US, by 1951, a new agency, the National
Science Foundation, had been created to protect and support both
fundamental research and higher education.

An important difference in the relationship between science and
society in the UK and in the US is in the way scientific research is
funded. In the UK, because of the autonomy given to the research
councils in accordance with the so-called Haldane Principle, 'a
vertical division developed between research councils and university science on the one hand, and the executive activities of government departments on the other' (Zuckerman, 1968). In the US, the
élite universities remained financially independent of government,
and were obliged to seek funds from private patrons. The larger
private foundations such as Ford and Rockefeller play a much more
substantial role in financing research in the US than their equivalents elsewhere. The universities thus maintained close links with
private industry and were prevented from developing that division

between 'pure' and 'applied science' and between research and development, which characterises UK and to a great extent Western European science.

In the UK concern about the lack of accountability of the research councils to Parliament, and thus to the general public, eventually prompted the government to commission the Rothschild Report (1972). This report recommended that a 'customer-contractor principle' should be applied to the allocation of research funds, and substantial amounts of annually allocated public funds were to be transferred to major 'customer departments'. The government accepted this basic principle, and this concept in administrative accountability is now operative. However, it seems doubtful if the customer-contractor principle will bridge the cleavage between science and society, as the bureaucratic structure of decision-making makes it difficult for parliamentary members to become cognisant with the necessary information and expertise needed for serious appraisal of science policy. Vig (1969) has pointed out that members of parliament are largely dependent for information on public and private reports, ministerial statements and replies to parliamentary questions.

American 'mission-oriented' research (that is, research related to some identified social need) has generated huge and expensive projects – has become 'big science'. For example, when the American government perceived space advance as a necessity (after the USSR launched Sputnik 1 in 1957), then funds made available for the space programme increased by a factor of seventeen in only ten years. It also fostered a new federal agency, NASA, and engendered the creation of an advisory post to the government on science and technology. The whole complex of American funds for research done by private corporations, and the military nature of most 'mission-oriented' research have prompted some authors to call the advanced industrial nations a 'techno-complex' or 'corporate state'. On his retirement, President Eisenhower cautioned Americans on the threat to democracy of what he called the 'industrial-military complex'. Supporting this line of argument C.W. Mills (1956) suggested that the US is controlled and manipulated by an upper stratum of leaders from three crucial sectors of society: (i) the large corporations; (ii) the military complex; and (iii) the political executive:

these hierarchies of state, and corporation, and army, constitute the means of power, and as such they are now of a consequence not before equated in human history.

Many thinkers consider that scientists reflect the nationalistic orientations of their respective governments. Indeed a schism developed in the international community of scientists because of national military knowledge: for some years after the First World War, German scientists and scientific organisations were excluded from participation in international congresses. Once the state began to sponsor substantial scientific research then the direction of much research, both pure and applied, became political. As Haberer (1972) writes:

> most of the tangible supports of science in terms of the allocations of social resources, public funds, manpower and training facilities, as well as policy directions, derive from the needs of the nation-state.

Leaders in each field of science increasingly act to protect their positions and the resources available to them by lobbying members of government, and thus obtaining sectional representation in the governmental process. This produces the intensification of group-bargaining activities, and further illustrates the compromise by contemporary scientists of the principle of neutrality. A value dilemma arises between, on the one hand, the idealised self-perception of the scientific community as disinterested and fraternal, with goals and the means of achieving them in essential harmony and, on the other hand, the fundamentally materialist and political imperatives of the institutions by whom the scientists are employed.

Salomon (1973) emphasises the dilemma of conscience of the contemporary scientist by comparing the trial of Galileo by the Church authorities with the case of Oppenheimer when he appeared before the Security Board hearing in 1954. Oppenheimer had been director at Los Alamos. Under his supervision a team of physicists had developed the atomic bomb, but he had been associated with certain known communists in the 1930s, and was also critical of further nuclear arms developments in the US. Salomon suggests that in Galileo's case science was on trial for its opposition to 'truth' as prescribed by the Church, whilst Oppenheimer as a scientist was challenged for his reluctance to serve the state. It is further suggested that the scientist's code of conduct in modern industrial society is forced to submit 'so easily to the imperative of the state' that scientists are 'alienated in their function from society' (Salomon, 1973).

A general picture of the scientific community emerges from our discussion. From a small membership and simple institutional networks with basically *laissez-faire* relationships, science is now, as Haberer (1972) has observed, 'central to the social and political order'. It is 'a leading social institution with a massive constituency, an elaborate division of labour, and complex institutional structures'.

5 Conclusions

We have argued that scientific and technological developments have had an enormous impact on the social systems of the industrial states. We have discussed some important theories of social change; in particular, the Marxist materialist conception of historical development, in which social and economic relationships are thought to correspond to the particular stage of a society's technology was contrasted with the functionalist approach. Marxist theorists and functionalists both make assumptions about human nature and the behaviour of people in society which largely determine their respective descriptions of the process of social change, and which colour their prescriptive assertions.

Regarding the political ideology of science and technology, we have explored the pervasive nature of the scientific world-view, and the concern of some radical thinkers with scientism and the ideology of industrialisation. Their case against science and technology as oppressive forces in society is difficult to refute. Modern nation-states are continually introducing new technologies which largely determine the division of labour, working conditions, and levels of manpower required in manufacturing and service industries. Ellul (1965), for example, suggests that *la technique* – techniques and technical value systems – becomes an end in itself, and that it will no longer be within people's capacity to control technological developments. His view is that it will take a dramatic increase in the consciousness of people in industrialised societies in order that this scientism can be seriously challenged and the implications of new technology adequately assessed and controlled. (We return to this theme in Part III.)

We have given brief consideration to science policy. In advanced industrial societies scientists and scientific institutions have been obliged to become 'politicalised' and to act as political pressure groups, in a similar way to other interest groups. Scientists and technologists have not achieved much *direct* political power: for example, in the late 1960s and early 1970s, in the US there were substantial cutbacks in research funding, despite the protests of the scientific community. Appraising the effectiveness of the British scientific pressure groups, Vig (1969) concluded that 'if scientists were represented in government, it was mainly in their capacity as civil servants, or as individual experts serving on advisory committees.'

There seems to be a good case at present for asserting the 'politicalisation of science' rather than the 'scientisation of politics'. Despite anxieties regarding technocracy expressed by many writers, contemporary scientists and technologists have not assumed the leadership and decision-making roles in government forecast for them. Indeed, there are many scientists and technologists who feel that little has changed since Babbage (1830) condemned English snobbery and social prejudices toward scientists and engineers.

Much concern in the 1980s is with military arms production by the industrialised nations and the escalation of nuclear weapons, and also with the use of nuclear power as an energy source. There is much pressure-group opposition to these developments, which is informing public opinion. Public opinion can influence government, particularly in the 'representative democracies'. An increasing number of scientists throughout the industrialised world are entering the dialogue on the 'social responsibility' of science. For citizen-groups to have a significant effect on the technology assessment process, those active in such groups should be as well-informed as possible about relevant complex scientific and technical matters. Coates (1975), who is active in organising grass-roots political opposition to nuclear arms stocks, suggests that, although the educational process is necessarily demanding, there is fortunately 'growing evidence that many groups and individuals are willing to make the kind of effort and sacrifice necessary to give new meaning to citizen participation.'

The political perspective is used in Part II below as an integral part of our analysis of the social effects of specific scientific activities and technologies. Questions of political control of science and technology are taken up again in Part III.

References

Armytage, W.H.G. (1965), *The Rise of The Technocrats* (London: Routledge & Kegan Paul).

Babbage, C. (1830), *Reflections on the Decline of Science in England and some of its Causes* (Farnborough: Gregg, 1969).

Burnham, J. (1941), *The Managerial Revolution* (New York: Day).

Coates, J.F. (1975), 'Why public participation is essential in technology', in *The Politics of Technology*, eds. G. Boyle *et al* (London: Longman).

Dickson, D. (1974), *Alternative Technology* (Glasgow: Fontana).

Durkheim, E. (1898), *The Division of Labour in Society* (London: Free Press, 1964).

Ellul, J. (1965), *The Technological Society* (London: Cape).

Galbraith, J.K. (1969), *The New Industrial State* (Harmondsworth: Penguin).

Gummett, P. (1980), *Scientists in Whitehall* (Manchester: Manchester University Press).

Haberer, J. (1972), 'Politicalization in science', *Science*, vol. 178, 17 November, pp. 713-24.

Habermas, J. (1971), *Towards a Rational Society* (London: Heinemann).

Hessen, B. (1931), 'The Social and economic roots of Newton's *Principia*', *Science at the Crossroads*, eds. N. Bukhairn *et al* (London: Cass, 1973).

Lewontin, R. and Levins, R. (1976), 'The problem of Lysenkoism', in *The Radicalisation of Science*, eds. H. and S. Rose, *op. cit.*, pp. 32-64

Lindberg, L.N. and Scheingold, S.A. (1970), *Europe's Would-be. Polity* (New Jersey: Prentice-Hall).

Marcuse, H. (1974), *One-Dimensional Man: The Ideology of industrial Society* (London: Routledge & Kegan Paul).

Marx, K. (1859), *Preface to 'The critique of political economy'*, in *Selected Works*, vol. 1 (Moscow: Progress).

Mills, C.W. (1956), *The Power Elite* (New York: Oxford University Press).

Partridge, P.H. (1963), 'Some notes on the concept of power', *Political Studies,* vol. XI, pp. 107-25.

Rose, H. and Rose, S. (1969), *Science and Society* (London: Allen Lane).

Rose, H. and Rose, S. eds. (1976), *The Radicalisation of Science* (London: Macmillan).

Rothschild Report (1972), *A Framework for Government Research & Development,* Cmnd 1272 (London: HMSO).

Salomon, J.J. (1973), *Science and Politics* (London: Macmillan).

Smelser, N.J. (1959), *Social Change in the Industrial Revolution* (London: Routledge & Kegan Paul).

Vig, N.J. (1969), *Science and Technology in British Politics* (Oxford: Pergamon Press).

Williams, R. (1971), *Politics and Technology* (London: Macmillan).

Zuckerman, S. (1968), 'Scientists in the arena', in *Decision-Making in National Science Policy,* ed. A. de Reuck, *et al.* (London:) Churchill).

5 Economic Perspectives

BRIAN STURGESS AND PETER WHEALE

1 Introduction

Economic theory comprises a number of competing schools of thought, and indeed some thinkers consider we are witnessing, because of lack of consensus, a crisis in economic theory (see, for example, Bell and Kristol, 1981). But economic analysis of society is of great practical importance and cannot be ignored just because its theories are controversial; indeed in many ways the application of science and technology in society *is* economics.

In this chapter the economic perspectives on science, technology and society are examined with the minimum of technical detail. In Section 2 the reader is guided through a simple exposition of neo-classical economic theory and its application to market analysis; and through cost-benefit analysis, in Section 3, to the problem of pollution control. Section 4 argues a critique of the assumptions of neo-classical economic theory and attempts to clarify some of the problems of technology assessment from an economic perspective. National economic policies also have importance for growth and employment, and so Section 5 includes a discussion of macroeconomic theory which briefly explains the controversy in this field and its policy implications for government. The chapter concludes with a consideration of the economics of innovation, which is so important to industrial development.

2 Neo-classical Economics

It was perhaps not until the English economist Marshall published *Principles of Economics* (1890) that economics as a discipline emulated science, separated itself from 'political economy' and became an amoral set of instrumental activities. Classical economists, from Adam Smith (1723 – 90) to John Stuart Mill (1806 – 73) were concerned with wealth, growth and economic welfare. The 'real' value of goods, they believed, was determined by the cost of production and not demand-price as neo-classical theory was to argue. Marshall's achievement was to develop an explanatory structure for market analysis which was eventually linked with the subsequent mathematical formalisation of 'general equilibrium' theory. Marshall brought together supply and demand curves which he suggested could be pictured like scissors intersecting. Figure 5.1 depicts this simple *static* relationship between the supply and demand of a commodity. The horizontal axis represents the amount supplied by the producer and the vertical axis represents the price of the commodity. The demand curve depicts the demand for the commodity by consumers and the supply curve the quantity suppliers are willing to supply at each price. Equilibrium is attained at point E where the supply and demand curves intersect; at this point the market for this commodity is cleared, at equilibrium price (p) and quantity (q).

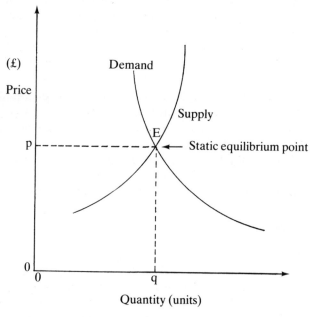

Figure 5.1: Demand and supply curves of a commodity

'Marginality' in the 1870s became a key concept in economics. The assumption was that the satisfaction or utility derived from a good declines as its consumption increases (i.e. an individual will only purchase one more unit of a good per period of time if its price is at least equal to the marginal utility derived from the extra unit of consumption).

In the 1930s, Robbins attempted to assess the scope of the new economics, and to define the characteristics of economic analysis. Economics had become the 'science' of allocating given quantities of scarce resources among competing claims, to obtain the most efficient or optimal use. His definition of economics is perhaps the clearest attempt to reveal the meaning of neo-classical economics. He stated that economics 'is the science which studies human behaviour as a relationship between ends and scarce means which have alternative uses.' The nature of 'economic man' is that of an agent who enters the world with particular objectives. 'Economic man', according to Marshall, was egoistically concerned with satisfying his own economic wants in a purely subjective way, his usual objective being the maximisation of his utility (i.e. his personal satisfaction, as a consumer; and in the case of a firm, its profitability). However, in attempting to attain these given objectives, the agent finds that the means of attaining them, such as income or resources, whether or not he has access to a relatively large or small amount, must always be scarce. Scarcity is thus defined in relation to ends and economic choices. It involves the notion of 'opportunity cost' which takes into account the next best alternative uses of means in relation to an individual's given objectives, when ends are many and means are few. As long as technology is not at a stage at which all wants can be satiated, then economics is an applicable mode of analysis. It is this interaction between wants and technology which, as we noted at the outset of this chapter, is at the core of economics.

Research and development require the employment of resources which have an 'opportunity cost', and so economics can be a useful form of analysis. Take, for example, a company wishing to produce for current profits in an industry in which technical change could affect future profits. It must decide how much of its resources it should devote to, say, advertising and production now, which are likely to increase current profits, and how much expenditure to devote to research and development which may increase future profits. In this case there is a trade-off between current and future values of the firm's objectives. Similarly, a government has to decide within its budget how much to devote to current expenditure for immediate needs, and how much to devote to expenditure which may provide future benefits.

Once a decision has been made which involves some knowledge of society's rate of time preference (that is, how present benefits are valued against future benefits), there remains the further complication of allocating scarce resources between, for example, medical, military or industrial-technological sectors. This implication of economic analysis has been highly developed, and has become known as 'cost-benefit analysis'. Cost-benefit analysis is often used as an aid to decision-making. It involves the need to place some value upon the benefits obtained from the competing uses of resources in order that comparisons of net cost-benefit can be made and decisions taken which maximise utility. The economist is thus concerned with the idea of optimisation (maximisation of benefits, minimisation of costs). This is the core of modern neo-classical economic theory.

3   Pollution Control

As an example of neo-classical economic analysis we consider the problem of pollution control. Pollution may be defined as undesirable changes to the environment, resulting from social or natural causes. The operation of the market mechanism (depicted in Figure 5.1) is quite unable to provide a socially optimal solution (known in economics as the 'Pareto optimum'), when one set of citizens wishes to dispose of waste or to deplete resources against the wishes of another set of citizens, who consider the consequences of such acts as detrimental to their environment. This problem also arises when the number of persons involved is too large to allow for a bargaining solution, or litigation is impracticable.

Cost-benefit analysis considers whether or not the total benefits of a polluting activity exceed the total costs of such an activity, a good example of which would be pollution caused by economic growth: the benefits of this pollutant activity are 'traded off' against the negative external costs. See Figure 5.2: the vertical axis represents both marginal (i.e. incremental) benefits and marginal costs measured in money terms, and the horizontal axis represents the rate of activities producing pollution. The socially optimal point is point A, at which the curves representing marginal cost to the environment and marginal benefit to the polluters intersect. At this point the marginal costs and benefits are given by the area OBAC, and the rate of economic growth causing pollution is OB. If polluters are allowed to expand production, exhausting all their marginal benefits at point E, then environmental costs to other members of society would be the area OEJQ. A similar analysis can be applied to any pollutant activity, though collecting measurable data and ascertaining the consequences to the environment often pose difficulties.

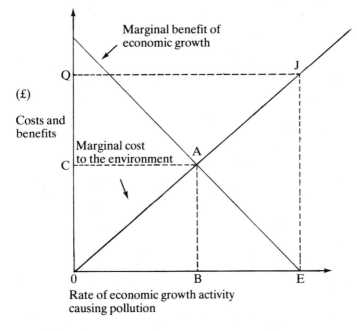

Figure 5.2: The costs and benefits of economic growth

Several proposals for pollution control have been made (for example, Rowley, 1974), and include public ownership, taxation, and legislation aimed at the control of pollution levels. Taxes may be levied on specific inputs and outputs in an attempt to attain the socially optimal level of pollution in cases where these inputs and outputs can be identified. Alternatively, pollution-intensive enterprises can be nationalised, and their plants then operated at the perceived optimal pollution rates either through a reduction in their rate of activity, or through the introduction of a cleaner technology. Imposing penalties on multinational companies can be expensive and cumbersome, but the international regulations on pollution control should be strengthened if the existing levels of pollution of air and sea are to be reduced. Governments in the West are susceptible to influence by the activities of pressure groups, and much depends on the success of the latter in persuading government that further pollution-prevention legislation is necessary at both a national and international level.

4   A Critique of Neo-classical Methodology

Governments tend to support high technology and potentially dangerous activities, such as the building of nuclear power stations which produce external effects such as pollution. A problem with using standard economic theory to analyse these issues is that it concentrates on *marginal* adjustments or changes in activity. Nuclear power, once decided upon, commits a vast amount of resources for a considerable time period, yet the pollution hazards associated with a major accident can hardly be said to impose marginal changes to the utility of affected individuals. Quite often the only redress of communities affected, or likely to be affected, by such developments is to engage in disturbance protests, or lobbying activities. Part of the problem is that the modes of analysis most often used in economics relate to these marginal adjustments or partial questions. It is easy for economists to answer questions about the change in the price of eye-glasses if the demand rises in a free market; but it is a much more difficult problem to calculate what happens to output and employment if the price of oil rises significantly, as it did in 1973. Thus models for analysing aspects of science and technology from an economic perspective should not rely only on marginal analysis, but need to take account of aggregate effects. There is a limit to the usefulness of disaggregated marginalist models when applied to many of the problems considered in Part II below. If we take as an example the analysis of industrialisation, many economic models are ahistorical and thus provide little guide when we attempt to unravel the processes which transform pre-industrial societies into the mass-consumption, high-technology economies we see today. Indeed, many economists who have attempted to explain industrialisation, or who today work on the problems of how to generate economic growth and industrialisation in less developed countries, are seen as being outside the main community of economists. Many mainstream economists seem unable or unwilling to consider such questions as: Does free trade or trade protection foster industrialisation and economic growth? Is growth compatible with a more equitable distrubution of income? Does high technology mean unemployment must inevitably increase? Despite the importance of these issues they remain of only peripheral interest to the majority of contemporary economists.

The immediate problems inherent in growth in modern industrialised economies have lessened with the contemporary adverse prospects for growth itself. Many of the present debates in macro-economics are concerned with more short-run issues, such as the control of inflation, the level of unemployment, and the ability of national economic policy to stabilise the economy. One source of difficulty is that the basis of theory in micro-economics (the inter-acting self-interested behaviour of people, households and firms) although possibly useful at the individual and market level for prediction and explanation, is inapplicable to the simple aggregated models which are the core of macro-economics. In most macro-economic models all consumer expenditure on the variety of goods available has to be aggregated into one variable at the macro-economic level. A similar aggregation process has to take place for labour and money. To be sure, models do exist illustrating the case where all markets are considered in the economy, and the conditions necessary for all to clear simultaneously were outlined by Walras as early as 1874. These models introduced the complex concept of 'general equilibrium' in the economy, but the failure of existing macro-economic models to predict actual economic responses to policy changes has led to a crisis in modern macro-economics.

5   Macro-Economics – A Paradigm Crisis?

The EEC countries and the USA are presently facing an economic recession almost paralleling that of the 1930s in terms of falling real output, declining living standards and steeply rising unemployment rates. There are differences, however. The 1930s saw a decline in the absolute level of prices, whereas inflation is a feature of the 1980s recession. This problem of inflation combined with recession has done much to create a crisis in macro-economic theory on a par with that prevailing in the 1930s, which later led to what has been termed the 'Keynesian revolution'. Keynes (1883 – 1946), in his *General Theory of Employment, Interest and Money* (1936), attacked the failure of the economic doctrines of the time to provide prescriptive measures to cope with the problems of world recession. Many of today's economists agree with Keynes that rigidities in money, wages and other prices are causal factors of lower levels of economic activity, though they still dispute the exact theoretical significance of his work. They discuss whether or not it was a general theory of the determination of the level of employment in capitalist economies; they argue about whether or not 'full employment' could be guaranteed by an unregulated market system, even if wages and prices were fully flexible, or if Keynes' work merely contained a more realistic analysis of economies and the policy measures necessary to maintain high levels of employment. Despite these debates on the nature of Keynes' contribution to economic theory, his policy conclusions concerning the management of the economy have been much used by post-war governments. The implication of the general theory is that governments could spend their way out of a depression by running a budget deficit financed by borrowing. A quasi-Keynesian policy was inadvertently used with some economic success in Nazi Germany; and increasing government expenditure had even been a major proposal of Lloyd George and the Liberal Party at the 1929 general election in Britain. An emphasis on public spending was also a major feature of Franklin Roosevelt's 'New Deal' policy in the USA. Indeed, it seemed by the late 1940s and 1950s that the classical school had almost admitted defeat in their assertion of the self-regulating nature of a market economy.

Full employment, it was widely believed, could not be assured without government intervention using the national budget to 'manage' aggregate demand. This became the avowed policy of successive British governments throughout the 1950s and 1960s, and a recognised policy of the US administration in the late 1960s and early 1970s. Emphasis was placed on the use of fiscal policy, government and taxation policies were emphasised, and monetary policies such as controls on lending, interest rates and the money supply were neglected. This neglect of monetary questions, although ironically not found in Keynes' work, became the hall-mark of the orthodox Keynesian school.

The growth of inflation in most developed countries in the late 1960s and 1970s formed the backdrop to a classical revival, criti-cising the Keynesian neglect of the rôle of money, a revival led by a group of economists who became known as 'monetarists', and inspired largely by the work of Friedman (1978). Monetarists believe the level of real output is invariable in the long run. As the stock of money in any particular period is determined by the government as a result of borrowing requirements, if the stock is increased, then for an economy in equilibrium, prices will rise. Therefore, accor-ding to the monetarists, governments' attempts to regulate the economy have resulted in raising income by increasing government expenditure, thus causing inflation. In the early 1980s many indus-trial countries including Britain and America began a monetarist experiment with their countries' economies, but while the rate of increase of inflation has been reduced, output remains low and unemployment high.

6 Economics of Innovation

It is, perhaps, above all else the application of scientific knowledge to industrial processes and production which has changed the way we work, eat and enjoy our leisure time. Technological innovations occur when an invention, or the embodiment of an invention in actual productive processes, is first used, and can be process- or product-oriented. Process innovations are those which affect the production of a particular product and are usually cost-reducing in that they save on some relatively expensive input into the produc-tion process. Labour-saving innovations are common and have complex social effects but, of course, not all innovations are labour-saving. When computerised banking systems were introduced in the

1960s, for example, the demand for labour in banking increased. Product innovations are a prominent feature of a modern economy, and they range from improvements on existing products, such as stereo systems and motor cars, to the introduction of products which are completely new, such as video recorders and personal computers.

From the viewpoint of the firm, innovations are aimed at reducing average costs by allowing greater output for the same cost, or by reducing total costs for the same output. In a simple framework of only two factors of production – capital and labour – such innovations can be characterised as capital-saving, labour-saving or neutral.

A strong link between innovation and economic growth has been suggested by many economists and historians, and various models have been proposed to attempt to explain the process of wealth creation from invention and innovation. The 'push' and 'pull' types of models postulate that market demand and high profits result from (are 'pushed' by) pure science, or that invention and innovation are engendered ('pulled') by some perceived social need. Various researchers have considered the time which elapses between invention, innovation, development, and the marketing of a commodity, but no definitive pattern has emerged. Mensch (1979), for example, has correlated innovations, which he identifies as fundamental or basic, with long-term economic cycles named after a Russian economist as 'Kondratieff cycles'. He concludes that long-term economic depressions and revivals are largely dependent on the availability of basic innovations. Mensch lists those innovations he considers basic, and his research indicates that they occur in clusters, and that these clusters precede long-run economic upturns in industrialised economies: 'We have abandoned the notion that the economy has developed in waves in favour of the theory that it has evolved through a series of intermittent innovative impulses' It is the frequency of inventions that provides the basis for basic industrial innovations. Historically, such clustering occurred in around 1825, 1875 and 1935.

Mensch's study of innovations is a further development of the interest in innovation itself, and in the 'dynamic efficiency' of the economy. This follows a long tradition in economics from Adam Smith's (1776) first statement of the 'invisible hand theorem' (that is, that a price system created out of the self-interest of traders and consumers guides an economy to static efficiency). Yet when there is a need to consider product and process innovations, efficiency must be measured in a dynamic sense, because resources, techniques and the goods which can be produced can change over time. Product and process innovations provide benefits, and because resources must be used up to provide these benefits, a more dynamic concept of efficiency is required.

One of the early students of the economics of innovation, Schumpeter (1942), stated:

A system – any system, economic or other – that at every given point in time fully utilises its possibilities to the best advantage may yet in the longer run be inferior to a system that does so at no given point of time, because the latter's failure to do so may be a condition for the level of speed of long-run performance.

Schumpeter described the 'gale of creative destruction' by which new products make old products obsolete and in turn themselves become obsolete as a result of the lure of profit. It is this dynamic process involving technology and innovation which the 'Austrian School' of economists argues is the real implication of the 'invisible hand' theorem. This School considers that the price system and the market system left to their own devices will always ensure dynamic efficiency. However, not all products which are desirable, and for which the technology exists, will be produced. Some products may be too risky for firms to produce alone (for example, Concorde and other aerospace innovations, or nuclear power stations) or may not be profitable for firms (e.g. some health products, and products for which the market is small).

Product and process innovations do not always produce benefits for all parties, and the need to develop an economic perspective in this book is based upon the question of control. In many cases the 'invisible hand theorem' does not provide the optimal solution. It may be that product innovations produce lower costs and greater output, but they may also lead to large-scale unemployment, to pollution, or other external costs. These are examples of what neo-classical economists term 'market failure'. Free-market mechanisms left to themselves either do not produce goods which are desirable, or produce goods by such means that external costs are imposed on other parties. As Arrow (1970), has argued:

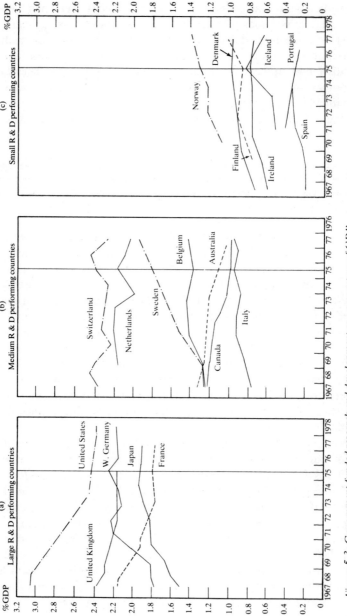

Figure 5.3: Government-funded research and development as a percentage of GDP.
Source: OECD Statistics, 1980.

A free enterprise economy will underinvest in invention and research (as compared with an ideal one) because it is risky, because the produce can be appropriated only to a limited extent.

The question of control leads to the question of what mechanisms society can use to harness the potential benefits of science and technology and to minimise the costs. The choice is between using market mechanisms or some form of state control; between state-controlled or state-regulated mechanisms, and private mechanisms. The rôle of government-funded research and development expenditure is large in many industrialised countries (Figure 5.3 gives government expenditures as a percentage of gross domestic product (GDP – the total value of the services and goods produced in a year) for large-performing, medium-performing and small-performing countries); state priorities are evidenced by the actual distribution of research and development expenditures. Since the early 1950s, the main priorities of public expenditures in the major industrialised nations have been on nuclear, military and space programmes. West Germany and Japan are exceptions because of their political positions after the Second World War. Table 5.1 gives some

Table 5.1: Percentage shares of government research and development expenditures for the United Kingdom and the United States by socio-economic objectives

	US			UK		
	1961	*1969*	*1979*	*1961*	*1969*	*1979*
Military, nuclear and space	89	79	62	80	59	58
Economic, agriculture and manufacturing	3	6	20	11	22	18
Welfare, health and environment	7	13	13	2	4	4
Other, including university	1	2	5	7	15	20

Source: OECD Statistics, 1980 (all figures rounded to nearest whole number).

statistics for the percentage shares of public research and development expenditures in the UK and the USA, and indicates the low priority given to welfare and environmental research and development programmes by comparison with the military and prestige types of science and technology.

Governments have been concerned to maintain 'consumer

sovereignty' and to enact anti-monopoly and consumer protection legislation to maintain consumer choice. Despite such legislation, the possibility cannot be ignored that in some cases, as Galbraith (1969) suggests, a 'producer sovereignty' exists where the innovators impose their preferences upon the consumer. The built-in obsolescence of many goods is a possible example of this.

There are, of course, many examples of consumer benefits from innovations in household, medical and communications products. Successful innovating firms must not only produce technically sound products, but must ensure that a market for their products actually exists. As Freeman (1974) writes: 'Most innovation case studies agree that those innovators who take considerable trouble to ascertain the future requirements of their customers are on the whole more successful.' The question of the incentive to innovate has spawned much debate in economics. Lack of investment opportunities in new technologies results, according to Mensch (1979), in only 'pseudo-innovations' to the existing products being made. These progressively less significant improvements are part of a process that manifests itself in the overall economic system as both stagnation and inflation. It is argued that stagnation is caused by the decline in demand for goods from the traditional sectors of the economy, and inflation from firms' attempts to bolster their positions by increasing prices, and their engagements in financial speculation, as a substitute for sound long-term innovation and new product development. Attention has also been concentrated on the type of economic system (or in the case of markets, market structure) which is most suited to meeting society's needs for new products and new means of producing products. If competition is desired, should all risk-reducing agreements be regarded as anti-competitive or efficiency-detracting? Many firms would not innovate if they were not enabled by patent legislation to obtain some limited monopoly on their new processes and products. There is a need to assess all activities and mechanisms by some criterion of efficiency, but this presupposes a dynamic definition of efficiency as yet not provided by economic theory.

It is questionable whether technical progress in aggregate has proved the blessing that it was thought to be. Consumers are not usually aware of the long-term side-effects of many products, nor of the aggregate consequences to society, yet the social costs incurred may often be greater than the aggregate benefits derived. (See for example Mishan, 1969).

A prediction of most neo-classical growth models is that, as *per capita* income rises as a result of innovation, then the marginal value of income to the consumer should fall, and that of leisure should rise. This would lead to a decrease in hours worked, and an increase in hours devoted to leisure activities. With the advent of labour-saving technology, the long-expected 'leisure revolution' was predicted. However, apart from the 'involuntary' leisure of unemployment due to economic recession, we seem further away from this 'revolution' than ever. In the US, hours of work and participation rates have stayed constant, or in some cases have actually increased as *per capita* income has risen. One explanation for this has been forwarded by a German economist Neumann (1976) using the distinction between product and process innovation discussed above. Whereas process innovation raises output per head of the population, and raises the marginal value of a 'unit' of leisure, product innovation (for example, in convenience foods and transport) designed to save consumer time often have the opposite effect. The net result, he argues, may be cancellation – people's leisure-work trade-off does not change, or may even lead to more work.

One interpretation of this phenomenon is that industrial society is in a never-ending treadmill, working more and more for more goods. This implies that, as long as well-being is believed to arise from an increasing consumption of goods, neither satisfaction, nor a 'leisure revolution' will be achieved. A fundamental change of attitude is required for the industrialised world to reject further material benefits – a shift, perhaps, towards a non-striving, Buddhist-like mentality. (For an exposition of this view, see Schumacher, 1974.)

We argued at the outset of this chapter that in many ways the application of science and technology in society *is* economics. Economic perspectives are therefore used on our analysis of the specific issues treated in Part II which follows.

References

Arrow, K.J. (1970), *Essays in the Theory of Risk-Bearing* (Amsterdam: North-Holland).

Bell, D. and Kristol, I. (1981) (eds.), *The Crisis in Economic Theory* (New York: Basic Books).

Friedman, M. (1978), *The Counter Revolution in Monetary Theory*, Occasional Paper 33 (London: IEA).

Freeman, C. (1974), *The Economics of Industrial Innovation* (Harmondsworth: Penguin).

Galbraith, J.K. (1969), *The New Industrial State* (Harmondsworth: Penguin).

Keynes, M. (1936), *The General Theory of Employment, Income and Money* (London: Macmillan).

Marshall, A. (1890), *Principles of Economics* (London).

Mensch, G. (1979), *Stalemate in Technology* (Cambridge, Mass.: Bollinger).

Mishan, E.J. (1969), *The Costs of Economic Growth* (Harmondsworth: Penguin).

Neumann, M. (1976), 'Innovation, Wachstum und Freizeit', *Kyklos,* vol. 29.

Robbins, L. (1935), *An Essay on the Nature and Significance of Economic Science* (London: Macmillan).

Rowley, C.K. (1974), 'Pollution and public theory', in Culyer, A.J., *Economic Policies and Social Goals* (London: Martin Robertson), pp. 284-312.

Schumpeter, J. (1942), *Capitalism, Socialism and Democracy* (New York: Harper Row).

Schumacher, E.F. (1974), *Small is Beautiful* (London: Sphere).

Smith, A. (1776), *An Inquiry into the Nature and Causes of the Wealth of Nations* (London), Books I-III, reprinted as *The Wealth of Nations* (Harmondsworth: Penguin, 1970).

Walras, L. (1874), *Elements of Pure Economics,* trans. William Jaffe, (London: Allen & Urwin, 1954).

Part II
KEY ISSUES IN ADVANCED
INDUSTRIAL SOCIETY

6   Food and Agriculture

PETER WHEALE

1   Introduction

Eating is essential for our survival, it is the source of our energy, protein, vitamins and essential minerals, and our evolution has interrelated closely with our diet. Technological developments have aided in providing our diet. A multitude of technological processes is now used in the food industry which convert raw food into edible food products and, by processing and preserving seasonal food which might otherwise be wasted, determine the general quantity and quality of food marketed. Science and changes in technology continue to increase the potential availability of food for human consumption.

However, criticisms are frequently heard of modern farming techniques, and of the food processing industries of industrialised nations. For example, the dangers of the use of pesticides and chemical fertilisers in food growing, and in the use of chemical additives in food processing, are emphasised. The food processing industry, some argue, is more concerned with the palatability of its products than with their nutritional value. National governments are accused of making inadequate attempts to ensure that their populaces enjoy healthy diets. Furthermore, despite increases in the productivity of food production, famines still occur in many less economically developed countries, with the resulting tragic consequences for the peoples of those countries.

In this chapter, we examine some notable technological innovations in agriculture in Section 2; the development and economic structure of the food processing industry are considered in Section 3; food consumption patterns are explored in relationship to social change in Section 4, and some illnesses associated with contemporary diet in the industrial countries are reviewed in Section 5. A number of important political issues related to food production is explored in Section 6. Finally, Section 7 deals with various economic aspects of food production, including agricultural efficiency, economic growth in the underdeveloped countries, ecological costs of modern food production, and concludes with a consideration of finance for the Third World, and the nature of its trading relationship with the industrialised countries.

2   Historical Developments

Prior to the Industrial Revolution the traditional system of husbandry was virtually self-sufficient: local resources largely served local needs. The farmer bred animal power, inherited agricultural lore and made most of his tools from local materials. Techniques such as pickling, salting and drying enabled people to eat fish and meat all the year round.

The introduction of new root crops such as swedes, turnips and mangolds, furnishing fodder for over-wintering cattle, greatly increased the efficiency of farming, particularly where the 'Norfolk four course' system (wheat, turnips, barley, clover) was adopted, thus avoiding the loss of production resulting from the need to leave land fallow. In the eighteenth century increased mechanisation on farms increased the output of farming in Europe and America. In Britain, the invention by Jethro Tull of the seed drill (1730) greatly increased agricultural productivity. Cast-iron ploughshares were manufactured about 1770 by Robert Ransome, and James Watt's steam engine of 1765 was later used to thresh and grind corn. By the early nineteenth century these innovations were in widespread use, not only in Britain but in much of Northern Europe and America. The newly independent United States was not only a willing recipient of these revolutionary farming machines but was soon providing her own agricultural innovations. Jethro Wood's iron plough (1814), and Cyrus McCormick's mechanised reaper a few years later, are examples of US innovations which greatly improved agricultural productivity.

During the early years of the Industrial Revolution in Britain, the practice of enclosing open fields was greatly extended, landowners preferring to lease their land to farmers rather than to numerous peasants as had previously been the case. A process of enclosing the open fields and common lands took place during this period on the

Continent also. At varying speeds (France, for example, was generally slower than Britain), the industrialising societies underwent a process of urbanisation in which there was a shift of labour from the rural areas to the cities and new towns. Consequently, the vast majority of people in these countries – approximately 80 per cent today – live miles from where food is actually grown.

British developments (accelerated enclosures and early industrialisation) made her agriculture particularly receptive to technological innovation, and this in turn led to a close relationship between farming and the newly developing industries in the towns and cities. Drainage engineers, chemists and fertiliser merchants began to play an important role in farming as they were eventually to do in all other industrial countries. The use of chemical fertilisers in particular was an important innovation, as Thomas (1979) argues:

The use of fertilisers completed the liberation of farmers from fallow. They no longer relied on cattle to provide manure, thereby breaking the age-old association between corn and cattle. That, in turn, freed farmers from their concern with forage, the cultivation of which had been essential since the beginning of agriculture.

In Britain, repeal of the Corn Laws (1846), which had placed a duty on imported corn, ushered in thirty years of prosperity for traders, whilst colonisation provided a secure source of cheap imports for Britain's developing industries. The number of farm implements during this 'Golden Age' of general prosperity increased dramatically and included reaping and threshing machines, iron ploughshares, harrows and cultivators, combine harvester-threshers, agricultural elevators for transferring and storing grain, and cheap wire (including the new barbed wire) for fencing. Early in the nineteenth century in France and England experiments in bottling fruit and the use of metal cans to preserve food were being carried out, and by the 1850s canning factories were marketing a range of soups and other food products. The new railways enabled the rapid transfer of fresh food from the country to the towns and cities.

Refrigerated meat imports in the 1880s from Australia, New Zealand and South America, grain from the United States brought in the new steam ships, and fruit from the East eventually brought depression to European agriculture; everywhere prices were forced down. Land was taken out of production, and lack of working capital for repairs and renewals of farm buildings and equipment resulted in a general run-down of agricultural production. Farm workers left the land in increasing numbers and industrial nations became far more dependent on imported food. (By the time of the First World War only one-quarter, for example, of Britain's food was home-grown.)

The inter-war years saw a continuing decline in European agriculture: in England and Wales, between 1871 and 1931, the total acreage under the plough fell from 14 million acres to 9½ million acres. In the United States, by 1920, reduced export demand for farm products ushered in two decades of depressed farm prices and incomes. The farmers themselves were short of working capital, and farms in need of repair and investment were simply allowed to degenerate. (See James, 1971.)

During this time, the industrialised nations' governments began taking interventionist steps to support their indigenous farmers. In the 1930s, marketing boards, especially for milk and potatoes, were set up to protect farming in Britain; import duties on food were imposed, and in certain cases crops were subsidised. In the US, a Federal farm board was established in 1929, empowered to assist farmers' marketing organisations, and to finance loans for the purchase of surplus commodities in order to raise prices. Although these measures brought some stability to the industry, farmers were not in a strong position to take advantage of them. In the US, insufficient funding and the weakened economy prevented government action of a magnitude sufficient to maintain farm prices, and the policies carried out as part of President Roosevelt's New Deal attack on the Great Depression did not prevent a shrinking demand for farm produce and the continued decline of farmers' incomes. (See Cochrange and Ryan, 1976.)

State intervention in the Second World War did provide a stimulus to agriculture: in North Europe and North America numbers of tractors in use trebled; combine harvesters increased thirty-fold; and land was reclaimed that had long since been taken out of production. After the war, legislation was enacted which guaranteed markets and assured prices; there were also state-financed drainage schemes. Mechanisation increased dramatically, but this tended further to reduce the numbers of workers employed in agriculture. Productivity gains were brought about as farmers increasingly substituted fertilisers, machinery, fuel, electricity, and other non-farm-produced inputs for human labour and land in food production.

Ownership and control, particularly since the war, have shifted from farmers to suppliers, distributors and processors. Suppliers provide fertilisers, pesticides and new crop varieties, whilst distributors assume marketing responsibility for agricultural production. The food processing industry is economically important; as a typical example Britain's food processing industry accounts for about 15 per cent of annual output, and is in receipt, together with agricultural commodities, of approximately 20 per cent of disposable consumer income. The experience of all the industrialised economies has been one of a movement of net resources out of agriculture into manufacturing and service industries. Increased agricultural productivity has been achieved by reducing manpower and increasing capital employed per man. Manpower in agriculture in the 1850s, in Britain, was about 20 per cent of the total workforce; by the 1920s it was under 10 per cent; and in the 1980s it is 2.5 per cent. For the US, figures for the same periods are approximately, 60, 20, and 5 percent; by contrast Nepal, for example, still has 93 percent of the workforce in agriculture. Similar, though less dramatic, trends ar eexhibited by all other economically developed countries: a tendency towards ever-more intensive, mechanised agriculture, and more single crop production – specialised farming or monoculture of one sort or another has become the norm.

3 The Food Processing Industry

Food technology is the application of food science to industrial food processes and, as we noted in Section 1, eating habits and food technology are closely linked. Until the mid-nineteenth century the food industry (manufacturing, wholesaling, retailing) was for the most part small-scale and supplied only local markets. Today, in the industrialised countries approximately 12.5 per cent of the workforce is engaged in the food industry and it accounts for around 20 per cent of consumers' disposable incomes. The industry's technology has become increasingly science-based and its structure highly concentrated. From a few hundred lines of processed foods available to the consumers in the 1950s, today there are in excess of 10,000 processed foods from which to choose.

When fresh foods are processed they undergo structural and chemical changes and some nutrients are usually lost. There is no truth, however, in the idea that when fresh foods are purchased and cooked all the nutrients are retained, whereas when the same foods pass through the hands of the food processor those nutrients are largely destroyed. Food processing covers many sorts of treatments carried out for a diversity of purposes: as examples, blanching, pasteurisation, sterilisation, dehydration, freezing, fermenting and cooking all cause some change in nutritional values. (See Bender, 1978.) Conditions for maximum nutrient retention are not always the same as those conferring maximum palatability, and manufacturers are sometimes accused of ignoring nutritional considerations.

In the industrialised nations there is no general evidence of malnutrition, and thus losses from food processing do not appear to be significant. Reduction of nutrients in foods may, however, take on significance when sections of the community limit their consumption to particular products: thus pensioners', babies' and very poor people's diets may, for this reason contain less than the recommended intake of nutrients. In the Third World there are the problems of relying on consumption of a limited range of processed products, a notorious example of which is the feeding to babies of manufactured powdered baby milk which has been diluted beyond adequate nutrient levels.

Another area of concern is additives to food-processed products: these additives (over 250 chemical substances are presently used) act as emulsifiers, stabilisers, antioxidants, artificial sweeteners, colouring agents, preservatives and solvents. The risk to health of consuming these chemical substances over a lifetime has been the subject of debate for some years now, but to date little is known about the combined effects upon health of long-term consumption of food products containing additives. But some evidence does exist that excessive intake of phosphates can lead to the premature cessation of bone growth in children, and that nitrates used in curing brines react with meat, producing small quantities of a known carcinogen. (See Wardle, 1977.) The concentrated structure of the food industry encourages the use of additives: for example, because the bread industry is highly centralised it needs to use antioxidants to help increase the shelf-life of bakery products and to prevent bulk fat used by the industry from going rancid.

4 Food and Social Change

Important social effects result from changes in food production techniques and in diet. Emphasising this, Reay Tannahill, in her book *Food in History* (1975), writes:

formerly, a man had been a hunter, not a herdsman, a plant-gatherer, not a grower. After the ice retreat, the deliberate culture of plants and domestication of livestock had begun, and the first villages had been established.

Humans early learned to be cautious about what they ate; Tudge (1979) estimates that primitive peoples eat perhaps only 10 per cent of the theoretically edible plants in their environment, and the culture and traditions of any society affect its consumption of different foods. Significant differences in food consumption patterns exist between the industrialised nations as well as between them and less developed nations.

Family structure and the level of nutritional knowledge also affect the pattern of food consumption. The process of industrialisation has affected the family as an institution in three basic ways: first, the centre of production has moved from the home to the factory; secondly, the factory has tended to employ individual workers rather than whole families, as had usually been the case with the cottage industries; and thirdly, the family has become a consumer unit. About three-quarters of the populations of industrial countries have an urbanised life-style resulting in a high consumption of processed foods; while reduction of average family size from six children in the nineteenth century to about two in the 1980s, together with economic growth, has enabled more *per capita* income to be spent on food in the industrialised countries.

In the West, more meat and vegetables, and proportionately less of staple foods such as bread and potatoes are eaten today than were eaten by our forebears. Many more married women work than did in 1900 (approximately a four-fold increase in the developed nations) and this is one reason for the increased demand for 'convenience foods' – 'convenience foods' are those manufactured to an advanced stage of preparation, thus requiring a shorter time to prepare in the home than less processed products.

While western urban dwellers are now able to enjoy an adequate diet this has not always been so, and despite increased affluence for farmers due to increased labour efficiency, conditions for many farm workers in the nineteenth century were not much improved. The consolidation of farms and the displacement of peasants from their holdings, the exaction of 'rack-rents', that is, rent equal to full annual value of property, and the curtailment of the commons, all contributed to the presence of poverty in Britain and on the Continent in the nineteenth century. Large numbers of the newly urbanised city dwellers experienced poor health due to malnutrition. Rickets and tuberculosis sapped their vitality, the first being a product of vitamin deficiency, the second intensified by poor diet.

5   Diet-Associated Illnesses

Limited understanding of the basic principles of human nutrition may explain why diet and health were not generally associated until the twentieth century. Scientists working throughout the world eventually showed that the health of animals declines when certain dietary components such as vitamins are missing. Von Liebig, in Germany, discovered that all herbivorous animals build up tissues from the proteins of plant foods which are converted into those of muscles and organs. The structure and composition of protein, the importance of amino acids in diet, and the nature of vitamins were not generally appreciated until around 1912. Serious diseases, it was then realised, can be caused by vitamin-deficient diets: lack of vitamin C produces scurvy, that of vitamin D can cause rickets, and that of vitamin A may lead to blindness. These vitamin-deficiency diseases have now been virtually eradicated from industrialised societies.

At the same time, the affluent societies have greatly increased the incidence of other diseases – tooth decay, peptic ulcers, gall stones, coronary failure, appendicitis, cancers of the bowel and urinary infections, and many others, as well as conditions such as constipation and obesity, all of which are thought to be the result of continuous consumption of over-refined foods and a high fat content diet. In particular, refined sugar, white flour and its products (from which the fibrous bran has been removed), and saturated fat (which, if eaten in excess, causes cholesterol to be deposited on arterial linings), are thought by most nutritionists to create serious health problems. The separation by manufacturers of nutritional value from palatability is partly responsible for this. Modern farming methods – the mechanisation and scientisation of agriculture – have also produced new health hazards for the farm worker. For example, farm workers risk the possibility of contracting lung diseases or cancers, and even poisoning from toxic substances, while operators of farm machinery risk osteoarthritis caused by machine vibration, as well as noise-related disabilities.

When processed food products and new technologies in farming are introduced into underdeveloped countries, the impact on traditional ways of life is dramatic. Lack of nutritional knowledge, the desire to imitate the West, and pressures from multinational corporations selling their products, often lead to tragic results for the consumers. Poor diets and illnesses result from loss of nutrients from processed foods when they are inappropriately prepared, or when western facilities are unavailable, as for example, where it is impossible for mothers to sterilise babies' feeding bottles. The world's population will be in excess of 6 billion by the year 2000 according to the United Nations Food and Agriculture Organisation (FAO) forecasts, and concern has been growing as to the fate of those hundred of millions (about 500 million in 1982) of people who are underfed. The physical and social effects of malnutrition are totally debilitating to people too poor to buy enough food for an adequate diet.

6 Politics of Food Production

Since the First World War, farmers have increasingly relied on government intervention in the markets for food products in Europe and America. Farm policy goals have generally been to maintain a prosperous and productive agricultural sector in the economies of the industrialised nations, but important differences arise with regard to the means of achieving these goals. Four basic politico-economic philosophies have been pursued at various times by the industrial nations:

1. A free-market philosophy, in which farm incomes and agricultural structure are determined by market forces.
2. A philosophy of 'equitable' farm incomes, achieved by govern-

ment subsidies to farmers where the operation of market prices fails to achieve this.
3. A 'protectionist' philosophy, in which either physical control on imports to a nation, or a system of tariffs on imports, effects an increase in the price of imported food stuffs, and thus acts to protect the demand for home-grown food commodities.
4. Control of supply of farm commodities, where either voluntary or imposed restrictions are applied.

No nation has applied any one of these policies, or a combination of them, consistently or systematically. In the nineteenth century, the general view was that government should not intervene directly in economic matters, and its role was simply considered to be one of assuring an equitably functioning market-place. During the inter-war years of economic crisis in Europe and America, the recognition (eventually supported by advocates of Keynesian economics) that market forces often create inequities and socio-economic ills, which responsible governments should not ignore, engendered a change of government attitude.

In the 1930s payments to farmers from governments were authorised both for soil conservation practices and as income supplements. Other programmes instituted to aid consumers also stimulated demand, including direct distribution of surplus food, the school meals programme, and free milk for young children, but in general, prior to the Second World War, *ad hoc* government involvement in agricultural production seldom sought to influence patterns of consumption and nutrition.

Involvement reached a peak during the two world wars, and especially during the second, when the food ministries in European countries controlled virtually all aspects of food supply and distribution – and therefore consumption – through rationing. A series of measures, including the provision of certain foods to children and expectant mothers, and statutory minimum nutritional content of some basic foods, (e.g. white bread 'fortified' with vitamins and minerals lost in the processing of flour) exercised a certain control over nutrient intake.

After the Second World War, the industrialised nations enacted legislation which gave their farmers some price supports – most countries preferring direct income payments to farmers when minimum set prices were not obtainable in the market-place.

Supply controls were also imposed on farmers to contain surpluses of production though these were never very popular with farmers themselves. Integrationalist philosophy in Europe, at the same time, spawned the Council for Mutual Economic Assistance (Comecon) in Eastern Europe and the EEC in Northern Europe. Comecon was formed in 1949 as part of the Soviet answer to the Marshall Plan, America's post-war European recovery programme. The European Economic Community (EEC) was formed in 1957 uncer the Treaty of Rome, and has been enlarged from its original six countries to ten at the time of writing. The EEC, or Common Market as it is commonly called, is now the biggest trading bloc in the world.

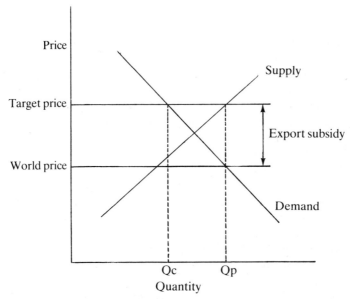

Figure 6.1: EEC price support system

EEC countries are bound by the Common Agricultural Policy (CAP). The common tariff barrier on foods entering from non-EEC countries means that there is sometimes a disparity between EEC and world food prices. The EEC sets a target price for each major product and imports are limited to the quantity which, with domestic supplies, can be sold at this price. The general trend to date has been for EEC prices to exceed world prices for most food commodities, encouraging farmers to produce surpluses which are supported by export subsidies borne by consumers within the EEC countries

Figure 6.1 shows the operation of this price support system, in which supply at EEC target price exceeds demand by QcQp. The unstored surpluses, which are subsidised by export subsidies to EEC farmers, are exported at world prices. Thus all the price effects of supply fluctuations are thrown onto the world market, further destabilising world food prices. In the Third World cheap food imports benefit relatively affluent city dwellers and put local farmers out of business (See Hill, 1975). Presently, over 60 per cent of the total EEC budget is allocated to the Common Agricultural Policy (CAP).

In the United States a system of farm income support was introduced by the 1973 Food and Agriculture Act. This system of supports is based on a deficiency payments mechanism (similar to that used in the UK prior to EEC membership). Under this system, farm prices are determined by supply and demand, and move up and down in line with world market price levels. In years when market prices fall below target prices set by the government, the government makes an income payment to producers. These deficiency payments are complemented by a 'set-aside' programme, which, when invoked, requires a producer to leave fallow a designated proportion of productive cropland if he is to be eligible for deficiency payments and loan support. This measure constrains the budgetary cost of the farm programme, and acts as a quota limitation on production when market prices are depressed. A farmer may exceed his quota by planting all of his available acreage, but must then accept the current world price for his crops. Unlike the European CAP, American agricultural policy ensures that American consumers pay only the world market price for food. By reacting to changes in world supply and demand conditions, the policy does not increase world market instability.

Economic rather than nutritional reasons are uppermost in the mind of central governments, and scant nutritional education is included in the general curriculum of children even though poorly balanced diets are thought to have adverse effects on health. Thus, this form of government non-intervention cannot easily be justified on the basis of preserving 'consumer sovereignty', particularly when governments are already subsidising producers.

Food manufacturing is largely controlled by multinational corporations producing many kinds of food products and often incorporating the production and retailing of their products. In most processed food markets two or three large firms dominate the market. The question arises as to whether the competitive forces are not too weak to ensure rapid diffusion of new techniques which could benefit consumers. Consumer pressure groups have far less political and economic power than the manufacturers' or the farmers' lobbying organisations.

There are bodies which deal with controlling food quality, the use of additives, and the labelling and advertising of food. Most of these bodies are statutory, but vary between industrial countries. For example, Britain has a voluntary producers' body to monitor and test chemicals used in agriculture, unlike the United States and other EEC countries where such bodies have statutory powers. Food standards and food additives bodies are government agencies, but inadequate laboratory facilities and small numbers of scientific staff often substantially reduce their effectiveness. Reliance is thus placed on the integrity of commercial staff in testing and monitoring food quality. Most of the advanced industrial nations are in the process of improving labelling regulations whereby manufacturers are obliged to give details of food content including additives. Is the public therefore protected? As Tudge (1979) suggests when discussing the US Food and Drug Administration, its task 'is not to provide people with alternatives to industrial products but merely to see that the only 'alternative' proffered is not immediately dangerous.'

The most marked contrast to the agricultural policy of western countries is that of the communist countries. After the 1917 Revolution, a policy of collectivisation of peasants' land was pursued by the USSR. Between 1929–33 this process was greatly accelerated in order to ensure a ready supply of agricultural products, to force the bulk of the Russian rural population into the towns, and to establish state direction of the land. This policy resulted in tens of thousands of Russians dying of starvation, and several millions being deported or fleeing to the cities.

In 1949, under the communist leadership of Mao Tse-tung, China began a co-operative movement, and from 1958 larger agricultural communes were organised, agricultural self-reliance being their objective. These collectives appear to have fared better than their Russian counterparts. Thus, although food rationing has been continuous, famine appears to have been averted.

Economic development is seen by many as the only way to eradicate poverty in the underdeveloped countries: others consider social and political change a prerequisite for economic growth and increases in *per capita* income. Western nations (the US in particular) have been reluctant to concede political changes in underdeveloped countries when these changes are of a left-wing nature. Examples abound of US intervention, direct and indirect, in the affairs of such countries: the constitutionally elected government of Guatemala in 1954 was overthrown by an invasion backed by the CIA and the United Fruit Company after the government, led by Arbenz, had begun a land reform programme involving the nationalisation of 200,000 acres of land owned by the United Fruit Company. In 1965, 20,000 US marines were sent by the US government to crush a rebellion in the Dominican Republic, led by Bosch who opposed US control, and might have cancelled molasses and sugar contracts negotiated with the US. In 1974 Allende, the Chilean leader, was murdered in a military *coup* backed, so his supporters believe, by CIA funds, and the new right-wing government led by General Pinochet was recognised by the US – Allende, a democratic Marxist, had wished to make sweeping land reforms and had nationalisation plans for US-owned interests in Chile!

North American exports of grain, rice and soybeans account for over half of the world's trade in these products, and there is concern about the political power accruing to those deciding, in grain-short years, who gets the grain and at what price: this aspect of grain production gives it a new political significance. Furthermore, the world's genetic pools of seed strains are being depleted as the new, uniform, high-yielding hybrid seeds, developed and distributed by multinational corporations, displace traditional varieties. Formerly, farmers stored seeds from the previous year's harvest to plant the following year, but as the hybrids do not always reproduce themselves, farmers often sell all their crop, and so need to go back each year to the seed companies for next year's seeds. United Brands, for example, now controls two-thirds of the banana genetic material; and there are similar collections for other crops. Shell-Oil markets seeds through three score companies in Europe,

North America and Africa.

The replacement of traditional crops by high-yield varieties – the so-called 'Green Revolution' – was originally financed by foundations such as Ford and Rockerfeller, and has generated a profitable world market for the world's chemical companies, dominated by such companies as Ciba-Geigy, Sandoz, Pfizer, Upjohn, Monsanto, Union Carbide and Royal Dutch Shell. It is perhaps not surprising that the type of hybrid seed developed requires expensive inputs such as the herbicides, pesticides and fungicides produced by these corporations, who also have plans to use the new techniques of biotechnology to culture protein-rich bacteria and algae in order to provide more food for the future (Yoxen, 1983).

Susan George, in her book *How the Other Half Dies* (1976), has argued that institutions such as the World Bank (IBRD), the associated International Finance Coporation (IFC), and the International Development Agency (IDA) only worsen the position of the Third World peasant farmer by encouraging the use of modern machinery and high fuel and chemical inputs. This leads to the growth of larger farms at the expense of small ones (modern machines work more efficiently with large areas of land), and to the displacement of labour. The end-result is the breakdown of local employment patterns, local food-crop production, consumer tastes and community life in the villages.

7   Economics of Agriculture

7.1   *Agricultural Efficiency*

Productivity has greatly increased in farming, but the increased use of fossil fuels as inputs to agricultural production has resulted in a proportionate decline in the energy output of major food crops measured in calories per unit of energy input. Crops and livestock form biological populations, and, like all living things, require a source of energy to survive. Farm livestock process plant materials to produce mainly meat and milk: the energy input – output ratio is about 7:1. The quantity of human food produced per acre of arable land is increased greatly by crop rather than animal production – an argument often used to support shifting from a meat-based diet to a vegetarian-oriented diet.

In economic terms efficiency merely relates input costs (fuel chemicals, land, labour) to the market price of produce as determined (in a free market) by consumer-demand tastes. The calorific value of output may be high when its market value is low because of consumer tastes. There are thus two fundamentally different rational criteria for assessing agricultural efficiency: a technical one, relating calorific inputs to calorific and nutritional outputs; and an economic one, relating the market prices of inputs and outputs of agricultural production. Governments may choose to control market value by applying taxes or subsidies to outputs (either to sellers or purchasers), but the economics of the markets for 'luxury' food products sustains farm production of meat and poultry and, in the Third World, cash crops for export which supply the foreign exchange needed for development. ('Luxury' food products have high income elasticities, that is, the quantities consumed are very responsive to income changes.)

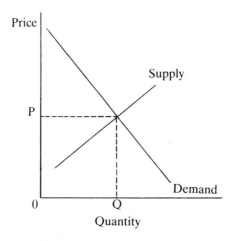

Figure 6.2: Equilibrium market price quantity

7.2   Change in Agricultural Technology
Given a freely determined market price, set under competitive conditions, the production capacity of the farm industry would equate the aggregate demand function of consumers as shown in Figure 6.2 At equilibrium price OP, total production would be OQ,

and the whole of OQ would be absorbed by the market. New technology inputs such as tractors, milking machines and chemical fertilisers and pest controls shift the supply function to the right and cause a fall in price (in the absence of any complementary shift in the demand function).

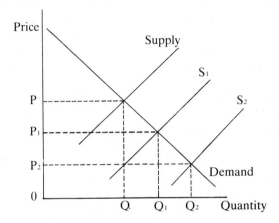

Figure 6.3: Effect of technology change on supply

Figure 6.3 shows this effect: S_1 and S_2 show the progressive effects of using new technologies which increase productivity, (i.e. the input-output ratio as measured by unit costs). In the absence of any complementary shift in the demand function, the price will fall and the supply coming on to the market will be wholly absorbed by consumers at equilibrium prices OP_1 and OP_2 respectively. At these prices the total supply OQ_1 and OQ_2 will be absorbed by the market. Effective demand in economics is defined as the willingness and ability to pay for goods; thus unless people can afford to purchase food at (in our examples) OP, OP_1, or OP_2 equilibrium prices, respectively, many will not be able to afford an adequate diet.

In market theory uninhibited trading, where each region produces those goods in which it has a 'comparative advantage' of some kind, leads to the best production/consumption of goods. Primary commodities (staple foods) have a lower consumer demand responsiveness to increased incomes (i.e. a low income elasticity) and as Third World countries mostly produce primary commodities, they tend to gain proportionately less economically from material progress created by secondary and tertiary industries, which are mainly based in the industrialised countries. Price is not determined in practice by a free market where equal access to information obtains: weather satellites and teletype networks assist the western nations to make faster and more knowledgeable decisions about conditions which may affect the future prices of food commodities. This information is invaluable when deals are made on the commodity future markets (the markets for buying and selling commodities to be delivered some time in the future).

7.3   Food Technology Transfer

The transfer of technology and the newly-developed, high-yield plant varieties have had only limited success in improving agricultural productivity in underdeveloped countries. Dickson (1974), whose work was referred to in Chapter 4 above, considers that technology transfer is an important means by which the industrialised countries maintain economically dominant relationships with developing countries. Once the disguise of neutrality is stripped away, Dickson believes, we see technology as a means of social control. According to Yoxen (1983)

> The developing ability to design, create and patent specific kinds of plants will confer upon the suppliers of plant varieties a greater degree of control over what is grown, over what substances are bought to protect or increase yields, over the price at which seeds are sold, and over the purpose for which crops are grown.

Alternative technologies have been advocated by a number of people and vary substantially in their technical content. However, they all aim to reduce the technology gap between imported equipment from the industrialised nations and the techniques of the recipient country. Sometimes an attempt is made to invest alternative technology with an alternative ideology to that of western capitalism: for example, Schumacher's (1974) 'intermediate technology' incorporates 'Buddhist economics' (i.e. low-cost, ecologically non-aggressive and co-operative production). In arguing for alternative technology Sardar and Rosser-Owen (1977) comment:

> Besides being labour-intensive, alternative technologies aim at being simpler and easier for people with a low level of technological culture to

understand and practise. The equipment involved is much cheaper and offers a low scale of production – a distinct advantage for small markets.

7.4   Stages of Economic Growth

In the 1960s it was believed by many economists that under-developed countries simply needed adequate capital to ensure their evolution into economically developed nations. Rostow (1960) argues that economies move through stages of economic develop-ment: starting with a traditional stage, they move to a 'take-off' period and once this is achieved, stages of sustained growth, maturity and ultimately high mass consumption occur. The implied assumption is that the underdeveloped nations are in analogous circumstances to those of the mature nations prior to industrialisa-tion. Rostow's 'stages of economic growth' theory however, ignores the economic relationship which has developed between the developed and the underdeveloped nations. Emphasising this his-torical relationship Kemp (1978) writes:

> The multilateral system of trade . . . enabled highly specialised economies to be brought together in the same system, but it was primarily a division between the industrialising countries and the primary producers, limiting the latter's development.

Developed countries were never in the past underdeveloped in the way Third World countries are today, because of the formers' continuing economic dominant relationship with the latter in inter-national trade. Rostow dismisses the problems of 'imperialism' as rhetoric, and twenty years after his book *Stages of Economic Growth*, he writes:

> the appropriate perspective on North–South relations is not only one of income redistribution from the rich to the poor, but of so maintaining the growth process in the world economy that the normal catching-up process operates on behalf of the latecomers (Rostow, 1980).

A form of economic dualism exists between the industrial and agricultural sectors of the developing nations, which results in un-balanced internal growth of the developing nations. Improvements in agricultural productivity are inadequate, as is the development of the infrastructure (transport, communications, banking, etc.) to match the requirements of a rapidly developing industrial economic sector. Political instability is closely related to unbalanced economic development.

In 1961, the Kennedy administration offered Latin America an economic and social 'Alliance for Progress' which involved long-term finance facilities. Chile, characteristic of developing countries, was a recipient of this foreign finance. But its agrarian sector is still feudalistic, whilst its industrial sector has developed rapidly. The Chilean economy is highly dependent on one primary commodity – copper – which has a volatile world price. As a result, Chile is now a massive food importer, and is dependent on continuing foreign loan finance. Social cleavages exist between country- and city-dwellers, and class conflicts exist between skilled workers and professionals, and the poor unskilled masses. It was the pressure of these social and economic cleavages, coupled with pressure from the US State Department using its influence to prevent the International Bank, International Monetary Fund, and the Inter-American Development-ment Bank from lending to Chile in 1972-3, which undermined the Allende government when it pursued extensive land reforms and nationalisation programmes.

7.5   Social Costs and Ecology

In economics those costs of production which are not borne by the producers, but which affect society, are called *social costs*. For example, it is society which pays, rather than the producers, for the cost of treating patients suffering from illnesses induced by exposure to pollutants. Agriculture involves alterations to the balance between the natural environment and plants and animals. Animal breeding, crop cultivation, forestry, dammed rivers, and human migration all change this balance (or eco-system). The social costs incurred by these various changes to the eco-system are very difficult to assess: for example, the social costs incurred by applying modern food cultivation techniques and food processing technology are impossible to quantify, especially as long time lags often occur before these costs become apparent. Concern about damage to the eco-system has intensified, however, in recent years, as more people consider that the balance between species based on their relationships in the food chain is being disrupted because of environmental damage. (See Carson, 1962.) The high-yielding plant varieties developed in the West require expensive inputs of fertilisers, irrigation, water and pesticides to sustain them. At first, it was thought by many that they would provide the answer to world starvation, but many problems associated with these new crop strains are now apparent, as Manning (1977) points out:

It has been claimed that the new strains lead to greater uniformity of crops and reduce the genetic pool. This, coupled with the trend towards mono-culture (large areas producing one crop), increases susceptibility to attack by pests. Chemical pesticides are expensive, tend to wipe out beneficial organisms as well as the target species and the reproductive capacity of pests is so great that resistant mutants rapidly appear on the scene. New chemicals must therefore constantly be developed, an increasingly expensive process, and the ecological side-effects of their use are often impossible to predict.

There are three types of pesticides used by farmers to control the organisms which reduce the quality or quantity of the famers' crop yield: herbicides, insecticides and fungicides. The argument against their extensive use is that the ecological principle of the eco-system is neglected. It is suggested that mercury-compound fungicides have seriously polluted water, and that organo-chlorine compounds such as DDT, Aldrin, Dieldrin, are persistent: i.e. once they have been applied their residues remain biologically potent for long time periods, thus threatening future crops and wild life. The use of these persistent products is now regulated in the developed countries.

Third World countries (where the majority of people are subsistence farmers) have been recipients of new plant varieties, fertilisers and pesticides, as well as mechanised farming equipment designed and manufactured in the industrialised countries. This has formed part of the 'technology transfer' process (i.e. the process of introducing new technology to an area, place, position or to a process, where it was not previously being applied). Not only has there been a number of ecological disasters in these countries as a result, but the problems of introducing advanced technology into alien cultures, where neither the funds nor expertise exist to evaluate the new techniques and integrate them with those already in existence, have often proved insurmountable. (See also 7.3 above.)

The US 'dust bowl', caused by damage to the soil structure due to over-intensive farming in the 1930s, is well known. Currently, the ecological balance is being disturbed, especially in regions of the Third World. Multinational corporations are encouraging changed land utilisation patterns by introducing the cultivation of industrial crops and luxury foods for export to rich countries. In Ethiopia, for example, cotton and coffee plantations have been expanded, displacing traditional pasture areas. In West Africa multinational companies profitably use thousands of acres for cash-crop farming, cotton growing and cattle ranching, at the expense of domestic grain production. In the Sahel region of the Sahara there has been overgrazing, deforestation, and diversion of water for cash crops, which have produced the advance of the desert and drought, resulting in recurrent famine.

The unforeseen side-effects of major irrigation schemes can be severe. The Aswan High Dam, for example, which dams the River Nile to provide irrigation for Egyptian farmers, caused serious soil erosion, a higher concentration of pollutants, and (because of increased salinity) wrecked the Egyptian sardine fisheries industry. In 1982, the Soviet Union embarked on a fresh-water supply project which aims to completely restructure the north Russian and Siberian river systems. It will submerge millions of acres of land and hundreds of historic monuments, resettle hundreds of thousands of Russians and probably disrupt the weather patterns of the whole northern hemisphere. It is feared that huge tracts of game forest will disappear, partially depleting Europe's oxygen supply which is partly manufactured by the plant and forest life of north Russia.

Agricultural innovations produced in the industrialised world, such as genetic improvement of crops and livestock by selective breeding, have revolutionised production. But in order to be economically efficient the farmer has increasingly specialised in crops and animal production. This has meant the abandoning of mixed farming, which previously provided manure for fertiliser, and crop rotation which preserved the soil fertility. The results of this new monoculture policy are huge increases in the inputs of inorganic fertilisers and pesticides.

Rothman (1972) has listed some less ecologically damaging methods of pest control, as follows:

1. Biological control: using the predators and parasites of pests.
2. Microbial control: using virus, bacterial and fungal diseases of pests.
3. Cultural control: using crop rotation, ploughing, crop diversification, and drainage controls.
4. Resistant crops: the development of genetic or nutritional techniques of increasing crop resistance to pests.

5. Autocidal control: by radiation or chemical sterilisation.
6. Interference techniques: using chemical attractants and repellants, radio frequencies and light frequencies.

7.6 Finance for the Third World

The Industrial Revolution in Europe was financed by private savings. As much as 60 per cent of firms' profits were ordinarily reinvested in the business enterprise in the nineteenth century. And British commercial and banking interests have invested huge amounts of funds in North American, Canadian and South African enterprises. But the underdeveloped countries have wholly inadequate funds generated internally, nor are the economic developments they wish to make often attractive to western financial institutions. (See Frank, 1981.) Foreign aid provides loans and sometimes technical expertise to underdeveloped nations, but loans are usually tied to imports from the source, to guarantee both raw materials to the developed nation in question, and a market for its manufactured products. And foreign 'experts' usually lack knowledge of the local cultural values and resources thus undermining the help they give. Also, unfortunately, among technicians sent by the United States and the Soviet Union have been officers of the KGB and the CIA which has discredited the work carried out.

Much investment in underdeveloped countries is to the advantage of the private multinationals, with little feedback in jobs and profits for the host-country. Considering US investment in Latin America, Easlea (1973) writes:

American manufacturing investment will increase the rate of production of commodity goods in underdeveloped countries but equally clearly American-owned manufacturing firms have no intention of competing with parent firms in the advanced capitalist countries. Thus private manufacturing investment in underdeveloped countries considerably limits their possibility of exporting manufactured goods in order to get the foreign currency much needed both for the purchase of foreign equipment as well as for payment of debt on past loans.

Investment funds are also obtained as foreign loans from the developed nations. A number of banking institutions lend funds to these countries, notably the International Bank for Reconstruction and Development (World Bank), the International Monetary Fund (IMF), and the International Finance Corporation (IFC). The results of the burden of debt are often a failure to satisfy the electorate's expectations and the destabilisation of the country's political system. Failure to raise the rate of capital formation has been exacerbated by the desire of developing nations to pursue prestige projects of various kinds and to aim at industrialisation rather than investing in agricultural development.

The North–South divide in food production and consumption is a result of the misallocation of resources on a world scale: the trade in cash crops for export by the underdeveloped nations, and the transfer of inappropriate technology to these countries by multi-national companies, are economically favourable to the West – the developed world, which contains only one-third of the world's people, uses half the world's supply of cereals, and feeds 70 per cent of its share to livestock; whilst the Third World, with two-thirds of the world's people, feeds only 10 per cent of its share of the world's cereals to animals. The report *North-South: A Programme for Survival* (1980), prepared by a Commission chaired by former West German Chancellor Willy Brandt, warns that the rich nations are threatened by the poverty of the developing world. This thoughtful report argues that the problems of monetary chaos and shortages of energy and raw materials, must be tackled on an international scale, and a global food programme developed and implemented. Reform of the world's financial system and large-scale transfers of development funds to poor countries are recommended. Several years have passed since the report's publication, and with the effects of world economic recession worsening, none of these recommendations has been implemented.

References

Bender, A. (1978), *Food Processing and Nutrition* (London: Academic Press).

Brandt Report (1980), *North-South: A Programme for Survival*, Report of the Independent Commission on International Development Issues (London: Pan).

Carson, R. (1962), *Silent Spring* (Harmondsworth: Penguin).

Cochrange, W.W. and Ryan, M.E. (1976), *American Farm Policy, 1948-1973* (Minneapolis: University of Minnesota Press).

Dickson, D. (1974), *Alternative Technology* (Glasgow: Fontana).

Easlea, B. (1973), *Liberation and the Aims of Science* (London: Chatto & Windus).

Frank, A.G. (1981), *Crisis in the Third World* (London: Heinemann).

George, S. (1976), *How the Other Half Dies* (Harmondsworth: Penguin).

Hill, B. (1975), *Britain's Agricultural Industry* (London: Heinemann).

James, G. (1971), *Agricultural Policy in Wealthy Countries* (London: Angus & Robertson).

Kemp, T. (1978), *Historical Patterns of Industrialization* (London: Longman).

Manning, D.H. (1977), *Society and Food* (London: Butterworths).

Rostow, W.W. (1960), *Stages of Economic Growth: A Non-Communist Manifesto* (Cambridge: Cambridge University Press).

Rostow, W.C. (1980), *Why the Poor Get Richer and the Rich Slow Down* (London: Macmillan).

Rothman, H. (1972), *Murderous Providence: A Study of Pollution in Industrial Societies* (London: Hart-Davis).

Sardar, Z. and Rosser-Owen, D.G. (1977), 'Science policy and developing countries', in *Science, Technology and Society,* ed. Spiegel-Rösing, I. and Solla Price, D. de (California: Sage).

Tannahill, R. (1975), *Food in History* (London: Paladin).

Thomas, H. (1979), *An Unfinished History of the World* (London: Hamish Hamilton; revised reprint, London: Pan, 1981).

Tudge, C. (1979), *The Famine Business* (Harmondsworth: Penguin).

Wardle, C. (1977), *Changing Food Habits in the UK* (London: ERRP).

Yoxen, E. (1983), *The Gene Business* (London: Pan).

7 Health and Medicine

PETER WHEALE

1 Introduction

Medical science and technology have undergone remarkable developments in the last century, and these changes have moulded a new social relationship between society and the medical profession. A new industry has emerged to supply the chemotherapeutic products which the developed world now demands: birth-control techniques, psychiatric medicine and high-technology diagnosis and surgery have had important social and economic effects on society. Developments such as legalised abortion, euthanasia and genetic engineering raise urgent moral and ethical questions.

In the economically underdeveloped countries, widespread disease, infant mortality and morbidity rates still pose enormous social problems, despite the application of modern medical knowledge and technology from the West.

In this chapter, Section 2 gives a brief outline of the most important developments in recent medical history which, taken together, amount to a 'scientific revolution' in medicine. Section 3 discusses the anxieties which have been expressed concerning certain aspects of social and political control achieved by medical intervention. Section 4 considers eugenics and the desires of the 'neo-Malthusians' to manipulate human evolution. Section 5 deals with the medical problems experienced in the Third World. Finally, Section 6 considers the concept of freedom of choice in the matter of health, and the problem of priorities in allocating scarce medical resources in a period of escalating costs.

2 The Rise of Modern Medicine

A 'scientific revolution' in medicine occurred in the nineteenth century. Two important innovations were the use of anaesthetics and antiseptics in surgery – by using ether or chloroform lengthy surgical operations could be performed; and antiseptics were used to prevent the infection of wounds during and after surgery. The English surgeon Joseph Lister (1827-1912), for example, used carbolic acid in 1865 to avoid infection in wounds during operations thus reducing the risks of septicaemia, erysipllas, tetanus and hospital gangrene. As a result, surgery was transformed.

The growth of hospital facilities and the beginnings of a nursing profession in the 1860s further helped to substantiate the revolution in medicine. But it was the advent of modern warfare which prompted the industrialised states to provide extensive casualty and medical services for troops stationed abroad, and to direct their attention to the health of civilians from whom they recruited. By the time of the Second World War it was obvious that the civilian population was likely to sustain serious injuries from enemy attacks, so requiring state-organised emergency services and central controls for health.

Prevention of disease by vaccination was another important early medical innovation. Edward Jenner (1749–1823) in 1796 proved that cowpox vaccine prevented the contraction of smallpox, after observing that the mild disease of cowpox imparted an acquired resistance to the often fatal disease of smallpox. His discovery led to the practice of vaccination against smallpox, and was widely used by the mid-nineteenth century. Louis Pasteur (1822–95) pioneered work on micro-organisms using improved microscopes and showed that the popular belief that micro-organisms generate spontaneously from dust, rotting meat or dung was mistaken. Causal organisms were recognised by Koch in 1883. Once a bacterial agent could be identified as the cause of a certain disease its control could be undertaken. Pasteur developed vaccines against rabies and anthrax. Virology is now a growth area of modern medicine. Salk's vaccine, for example, was developed at the University of Pittsburgh, and prevents poliomyelitis, immunisation having been introduced in 1956. Immunisation against tetanus, whooping cough, diphtheria and tuberculosis is now available in all the developed countries.

The development of this new medical knowledge, and the professionalisation of the medical establishment, transformed the social relationship between doctors and their patients. This professionalisation, embodied in medical councils backed by the state, ensured that it was doctors who decided on the criteria for entry into the medical profession, and applied the laws and codes of medical practice.

The turn of the century saw the identification of hormones and vitamins. The nutritional principle maintained that a vitamin deficiency caused by inadequate diet or the malfunction of certain organs could be corrected by vitamin supplements. The hormones adrenalin, thyroxin and insulin were isolated and studied, and the application of artificially synthesised hormones to hormone-deficient patients increased the life prospects for many suffering from debilitating diseases such as diabetes.

Chemotherapy began in earnest with Paul Ehrlich's 'magic bullet' Salvarsan for the treatment of syphilis in 1910. In the 1930s the development of the sulphonamide drugs in France and Britain proved effective in removing bacterial sepsis from childbirth, physical injury, and surgical operations.

The production of penicillin early in the Second World War opened the way for the development of modern antibiotics which replaced the sulphonamide drugs; in 1944, for example, strep-tomycin was used to treat tuberculosis, causing a fall in infant mortality. By 1938 the sex hormones, including progesterone, had been isolated in pure form and it was demonstrated that progesterone could be used to prevent ovulation in various female animals. A synthetic compound was developed that had a similar effect, but in addition was effective at low doses and could be taken orally. It took another twenty years, however, before both medical and public opinion were predisposed to the manufacture and dis-tribution of the 'pill'. Once considered respectable by the medical profession, the huge market for such contraceptive techniques was exploited by the large pharmaceutical companies (for example, G.D. Searle). This development has helped transform a whole generation's attitude towards sexual relationships. Unfortunately, though, this sexual revolution has been associated with increased incidences of venereal diseases and conditions such as herpes.

Scientific synthetic chemistry, practised in an economically con-centrated industry, has spawned a plethora of new drugs: chemicals with great potential for creating physiological and psychological change are now manufactured, and many are now concerned that the medical profession's central control over medication is being displaced by that of the modern drug industry (see, for example, Klass, 1975). Recognising a huge potential market, the new science-based industries in the last quarter century have produced scores of important medical innovations: artificial kidney machines, which filter out toxic impurities from the blood by dialysis and enable victims of chronic kidney disease to live active lives; the electro-cardiograph, the electroencephalograph and the electromyograph, to trace the electrical activities of body tissues for use in clinical diagnosis; the 'body scanner', was developed in the 1970s by EMI, which uses X-rays to obtain three-dimensional photographs of the patient's anatomy. (GEC have recently marketed a scanner which uses nuclear magnetic resonance (NMR), and not X-rays, to produce the required images. This equipment, it is claimed, provides greater diagnostic information, particularly for soft tissue and blood.)

Diagnostic techniques have been developed which introduce radioactive isotopes into the body. These are then monitored through the bloodstream by a Geiger-counter, or Gamma camera, so that obstructions caused by tumours or malfunctions can be located and examined. Cancers, brain damage and disease, some virus infections, and blood diseases can be diagnosed in this way. The discovery of metals and polymers suitable for use in ortho-paedic and arterial surgery, surgical lasers, developments in haematology and immunology, and transplantation of organs, are all important advances in medical treatment. Computer diagnosis, 'magnetic' scanners, statistical analysis, and data storage and retrieval systems have also been introduced in recent years. One commentator, Norton (1973) notes:

Automatic analysers and sequential multiple analysis have come just in time, it seems, to save clinical biochemistry laboratories from breakdown under mounting demands from clinical medicine – demands for the estimation of more and more constituents of blood, urine and other body fluids.

In order to deploy these new techniques, medical institutions must now employ large numbers of scientists and technicians from a wide range of disciplines. Of great importance and public concern in the 1980s is the expansion of new biological techniques which have only become possible with an appreciation of the structure and function-ing of cells. Robert Hooke (1635 – 1703), the English microscopist, had named cells in 1665 after he detected their dead remains in a shaving of cork; in 1673 Anton van Leeuwenhoek (1632 – 1723), a Dutchman, was the first to describe a living cell. Gregor Mendel (1822 – 84) crossbred garden peas in the garden of his Austrian monastery. His results, published in 1865, postulated the theory of inheritance through 'hereditary units', but his ideas were ahead of the cellular knowledge of the time and were ignored for the next thirty-five years. It was only in 1900, after the postulation of the connection between heredity and the 'bodies' (chromosomes) in the most prominent internal feature of the cell, the nucleus, that the importance of Mendel's discoveries was understood. It was then that the intensive study of heredity began, a discipline given the name of genetics. Mendel's 'hereditary unit' was called the gene and chemically identified as deoxyribonucleic acid (DNA). Encoded in the DNA are the 'instructions' for all the cellular processes and the reproduction of the cell.

In 1952 two experimental embryologists, Briggs and King, working at the Carnegie Institute in Washington, devised the technique of transferring a nucleus with its full complement of genetic information from one cell to another. In 1953, Watson and Crick, working in Cambridge, England, published a short paper describing the structure of DNA. It was the cracking of the genetic 'code' that heralded genetic engineering as a real possibility. Since the early 1950s biologists have made revolutionary progress: they have built genes artificially, and have transplanted genes from frogs, mice and fruit flies into bacteria. Many scientists believe that soon gene manipulation may be used to cure diseases as disparate as diabetes and cystic fibrosis. It may be used to cure cancer, prolong life, and create new hybrid plants and animals. Eventually, it may even be possible to produce limitless copies of people (or clones) by transplanting their DNA into human eggs!

By the early 1980s several technically successful genetic engineering products were available. Thus human growth hormone, insulin and the natural defence molecule interferon (which may be active against viruses such as those responsible for 'flu and polio, and some forms of cancer) have been developed in the laboratory (Yoxen, 1983). An oil-eating bacterium (*Pseudomonas*), which digests particular hydrocarbons of crude oil, has been developed in General Electric's research centre. Hundreds more products are currently being manufactured using the new biotechnology, including powerful analgesics, vaccines, drugs, detergents, plants, and plastics. Molecular biochemistry and microelectronics have recently combined to produce the 'biochip' – a combination of organic materials and microelectronics which generate artificial intelligence systems – in 1982 a US firm, EMV, patented a simple biochip. The implications of this innovation are that genetic engineering may be used to construct genes which are able to instruct a human cell to build microcomputers out of proteins. The creation of organic – electronic brains is theoretically possible – the robots of the future may be capable of internal growth, self-programming and organic change in the light of experience. These various developments pose enormous ethical problems, and raise questions of how and to what extent the state should control them.

3 Scientific Medicine and Social Control

From the mid-nineteenth century, medical science has been held in high esteem, and the doctors and researchers who practise it have become the main arbiters of the health of industrial societies. There has been a triumph of the scientific expert over the older medicine, symbolised by 'the frock-coat, the stock, and the wing collar'. 'White-coat' medicine means hospital medicine, and other branches such as general practice, preventive medicine and community medicine have been relegated to a lower status (Norton, 1973). The view of the scientific expert is that the subject-matter of medical science is the body. The machine serves as his model for living organisms, organs of the body being analogous to machine parts which sometimes need to be repaired to prevent the body (machine) from failing. Medical science has many of the characteristics of natural science: its language is the language of universals, and the individual is seen as a particular example of a universal aspect of nature. In health care the effect of this is that the patient is treated as an object – the material of medical science. The medical profession's attitude to women in labour illustrates this medical scientism: the woman is perceived as the subject-matter the hospital staff expect, and to numerous gynaecologists the patient's womb is simply the organ to bear children, removable if it impairs the overall efficient functioning of the human machine. (See for example, Morris, 1960.)

Lederman (1970) points out that patients are studied using different medical sciences, each with a deterministic approach (a 'mechanical materialist' approach). There is a tendency to conceive of patients' illnesses as the result of factors beyond their control. It is argued that by not respecting patients' freedom, a doctor conveys the feeling of inevitability. Lederman (1970) suggests that this leads to despondency and a sense of hopeiessness in those cases where the doctor cannot remove the cause of the disease or disability, and, as a result, patients do not attempt to help themselves.

The medical profession now forms an élite group which has supreme authority to pronounce upon health matters. The populations of industrialised countries mostly abdicate responsibility for their health, believing that the treatment of illness and disease is a matter only for trained doctors. Emphasising this point Lederman writes:

> The population swallow pills, submit to injections and to operations without having any idea what is happening to their bodies and minds under such treatment. They are the objects of deterministic science and not free people using their own intelligence.

Illich (1975) has analysed modern medicine and considered its effects on people's motivations and their potential to control their own lives. On balance, Illich argues, the medical system, including hospitals and doctors, actually damages our general health. He points to addiction of people to medical care as a solution to all their problems, and the delegation to doctors of the responsibility for health care (the conversion of social and political ills into 'illnesses'). Illich proposes a utopian solution to these problems – a deindustrialised and less bureaucratic society where individuals can choose their life-styles and regain their autonomy. He rejects the environmental or 'holistic' approach to medicine in which people's whole way of life is scrutinised, with the emphasis on preventive medicine, seeing this as leading potentially to an even greater control and manipulation, what he calls 'total treatment', of the individual by the medical profession. Whilst providing worthwhile insights into social effects of scientific medicine, Illich's analysis studiously underestimates the positive benefits of modern medicine (for example, the contributions to our well-being of anaesthetics and immunology). Furthermore, he avoids proper consideration of the underpinning economic structure of the advanced industrialised economies. In rejecting 'holistic' medicine, Illich rejects what many regard as a basic principle of good health. The radical view is that the prevention of illness depends on the conditions in which people conduct their lives, and only by considering the whole person can the prevention of, and treatment for, disease be fully achieved. The economic and political forces which undermine the individual's autonomy should be countered by political action (for example, pressure-group activity and 'counter-information'), but will certainly not go away by merely prescribing a Rousseauesque society.

The layman's view is that medical science has eradicated many formerly fatal diseases, but as McKeown (1979) has so convincingly shown, the impact of medical advances has been much smaller than is generally believed. Major infections such as tuberculosis, pneumonia, cholera and measles were declining long before successful intervention was possible and, generally speaking, immunisation and medication were less effective than other influences such as improved nutrition, housing and sanitation. At the end of his analysis of the role of medicine, McKeown warns us to beware of the certainty ascribed to the determination and elimination of infectious disease:

> the influences which determine man's response to infectious disease – genetic, nutritional, environmental and behavioural, as well as medical – are infinitely complex, and we need to be very cautious before assuming that we fully understand the infections, or that we have in our hands the certain means of their control.

The expansion of trade and demographic changes engendered by the Industrial Revolution encouraged the spread of infectious diseases, especially those derived from the transmission of micro-organisms that were air-, water- or food-borne. The only specific contribution of medicine at this time was the widespread use of vaccination against smallpox, a technique long preceding knowledge of immunology. Whilst acknowledging the beneficial contribution to health of chemotherapy in recent years in further reducing the incidence of many diseases (tuberculosis, diphtheria and polio, for example), we must admit that diseases of the advanced industrial society have appeared. Major causes of mortality now include cardiovascular conditions, coronary heart disease and high blood pressure, chronic infections of the respiratory tract, rheumatism, industrial accidents, congential deformities and mental disorders.

It is because medicine is perceived as a science producing objective knowledge and administered by 'experts' that it can so easily be used as a means of social control. The American sociologist Talcott Parsons (1972) recognised this when he considered the nature of the 'sick role'. People, he argued, who occupy this role are not held responsible for their incapacity and are therefore exempted from their usual obligations. However, the 'social exchange' is that they must want to get well and are thus obliged to seek and comply with appropriate medical advice. Anyone who deviates from the social norm risks being diagnosed as suffering from some form of illness. Threats to the social order can be eliminated or greatly lessened by forcing the 'deviant' to adopt the sick role. Medicine is thus a moral enterprise, like law and religion, seeking to uncover and control what it considers undesirable. What had been called 'crime' in the past – lunacy, degeneracy, sin, and even poverty – is now often called 'illness', and social policy has been moving toward adopting a perspective appropriate to the imputation of illness (Friedson, 1975). Psychiatric diagnosis is sometimes used to silence political opposition to the state: in the Soviet Union 'treatment' objectives include improving the political philosophy and value-systems of the 'patient' (See, for example, Clare, 1976.) In the nineteenth century, black slaves who tried to escape from their masters were said to be suffering from a disease called 'drapetomania', and those who refused to work were considered ill with 'dysaesthesia Aethiopis'. Berke (1979) gives this as an example of a medical metaphor being used in order to make moral, social or political judgements seem respectable.

Diagnosis of mental illness, and its treatment by chemotherapy, electro-convulsive treatment (ECT) or lobotomy (surgical removal of part of the brain) are particularly disturbing, especially since at least 10 per cent of the populations of the developed countries will spend some time in a mental hospital. Mental health enactments in all the developed countries give power to the authorities (effectively doctors) to restrain and institutionalise anyone they diagnose as mentally ill. Once institutionalised, patients suffer a process of depersonalisation and lose their citizen's rights, the restrictions varying with the degree to which they have become social nuisances. Some suggest that the psychiatric profession, which dominates decision-making in the field of mental health, is firmly rooted in the traditions of scientific medicine and that the methods of treatment adopted have little proven efficacy, and at best can only be considered palliative (Baruch and Treacher, 1978). Psychological treatments tend to be denigrated by the psychiatric profession, but they at least involve the establishing of human relationships, and thus are basically different from physical treatments, as relationships between people are themselves expressions of freedom.

The medical profession is often called on to act as the arbiter of life and death because scarce medical resources oblige doctors to discriminate between patients through the allocation of these resources. This professional autonomy is questionable where the decisions are based on economic and moral grounds rather than on technical ones. In 1981 the acquittal of a doctor in England for the attempted murder of a Down's Syndrome (mongol) baby highlighted the power over life and death wielded (albeit in a well-intentioned way) by the medical profession. The doctor, with the agreement of the baby's parents, had allowed him to die, applying a philosophy of 'non-treatment', a form of euthanasia condoned by the medical profession. (See Gillie, 1981.)

One area of medical science, already discussed, where the medical profession has *lost* effective supervision is that of drug manufacture. The developed countries exhibit a growing dependence upon drugs, in particular, an increasing dependence upon mood-elevating drugs for treating depression, as well as an enormous consumption of sedative drugs for tension relief. The contraceptive pill is another widely used drug in the West, and concern has been expressed about its safety. Evidence exists that this form of chemical contraception may, at least, 'unmask' cancer, diabetes or thrombo-embolic disease (Mintz, 1969, Vessey, 1978, Hawkins and Elden 1979). No statutory post-marketing surveillance system exists for gathering data on adverse drug reactions in the UK and the Committee on the Safety of Medicines (CSM) must rely on data submitted to it by the drug firms, and on medical publications.

Another important field of development over which the medical profession has little control is biotechnology. As in the manufacture of drugs, most research and development is 'mission-oriented' and profit-motivated; the personnel concerned are increasingly drawn from outside the medical profession. Official codes of conduct have been drawn up to protect society from potentially risky activities undertaken by scientists and commercial enterprises, for example, laboratory standards for quality control of drug manufacture are embodied in the United States Pharmacopoeia (USP), the National Formulary (NF), and the British Pharmacopoeia (BP) and activities in the field of genetic engineering were constrained in the UK by the now disbanded Genetic Manipulation Advisory Group (GMAG) (Yoxen, 1983). Public anxiety lies not in the inadequacy of these standards, but in the impossibility of enforcing them adequately. In the opinion of Klass (1975) it is the failure of the medical profession to continue critical supervision over the 'industrial–medical complex' which is undermining our health systems.

4   Eugenics and the Neo-Malthusians

In 1883 Sir Francis Galton, Charles Darwin's cousin, coined the term 'eugenics' to describe processes leading to the genetic improvement of the human species through selective breeding. Charles Darwin's (1859) theory of evolution, that natural selection was the process that gave direction to evolution, provided impetus for modern eugenics, both as a science and as a social movement. It should, however, be noted that he disagreed with preventive checks to populations on the basis that over-multiplication was useful, since it causes a struggle for existence in which only the strongest and the ablest survive; and he doubted whether it was possible for preventative checks to serve as effectively as selective forces. Taken in its broadest sense, eugenics incorporates the study of the genetic consequences of implementing policies of sterilisation, artificial insemination, population control practices, pre-natal diagnosis, and selective induced abortion, as well as the possibilities of biological manipulation using genetic engineering. As a social movement, eugenics encompasses all efforts whose goal is the modification of natural selection to bring about genetic change in a particular direction (to 'check the birth-rate of the unfit') within given human populations, or the human species as a whole. In the nineteenth century many social reformers held the view that birth control designed to achieve eugenic goals, should be a central part of social policies. These neo-Malthusians (so-called because their views were derived from the theory of population propounded by Malthus, 1798), believed that a eugenic distribution of births should be achieved primarily through 'negative' eugenics – social programmes designed to reduce the birth-rate of the less fit members of the population.

It is interesting that this view that the environment of nineteenth-century industrial society was producing selection that was dysgenic (i.e. leading to genetic deterioration), rather than eugenic, contradicted the view of the Social-Darwinists. The Social-Darwinists equated the struggle for existence with individualistic struggle within an economic *laissez-faire* society, and maintained that biological success and economic success coincided in such a way that the people who won wealth, social prestige and power were also the biologically fit. (From an evolutionary point of view, the biologically fit in a population are merely those who contribute the most genes to the next generation).

Modern industrial societies have, for the most part, accepted the view that birth control should be practised, believing that the regulation of the size as well as the genetic quality of human populations in a humane way improves the quality of life. The Malthusian League, the Birth Control League, the International Planned Parenthood Federation and other groups campaigned for the 'birth control movement', but attitudes concerning contraception were slow to change, and a favourable attitude to birth control did not occur until about the time when concern about a 'population explosion' became topical in the 1950s. Pincus (1975), who began his research for an oral contraceptive in 1951, cited the population explosion as an impetus to such research. The popular emotive association of the work of the birth control movement with the earlier eugenics movement had previously, and unfairly, discredited it in the eyes of the Church and of many lay groups and doctors who were concerned about quackery and dishonesty. Today, few would wish to associate the birth control bodies with eugenics, and the Family Planning Association in Britain, and the Planned Parenthood Federation in the US, are respectable organisations working with the medical profession.

Birth control technology has given enormous powers of regulation to the state; governments have policies to influence the social structure of their populations (for example, through child tax allowances, availability of contraceptives, abortion, and incentives for sterilisation). Knowledge of genetics and the accompanying medical technology now give prospective parents the information they need to control the genetic quality of their children (by selective abortion of foetuses known, or believed to be, carrying a defective chromosome).

Artificial insemination with donor sperm (AID), also contributes to the ability of governments and individuals to modify and control natural selection. Some have suggested that human beings should use artificial insemination with sperm of donors chosen on eugenic criteria. Compulsory sterilisation of groups of people, such as the mentally retarded, mentally ill and physically defective, has often been advocated as a means by which eugenic goals could be achieved, and compulsory sterilisation laws, for example, were passed in the United States during the early part of the twentieth century (Bajema, 1976). State control was at its most extreme and malign in Nazi Germany. In attempts to 'rationally modify' the future of the German race, at least 200,000 compulsory sterilisations were performed between 1933-45 under the Eugenic Sterilisation Act (1933). Euthanasia and genocide policies towards so-called 'inferior breeds', such as Jews and gypsies, were carried out on a scale unprecedented in history. After the Nuremburg trials of Nazi war criminals in 1946, the judicial tribunal, with the help of doctors, drafted an ethical code of conduct – the Nuremburg Code – regulating the relationship between experimenter and experimental subject.

Grotesque racist experiments were also conducted in Japan during the Second World War. Morimura and Shimozato (1982) have described how Japanese researchers attempted to determine the most effective biological and chemical agents for use against different races. Several thousand Allied prisoners died at the hands of the Japanese from these experiments which are purported to have included injections of plague, cholera, typhus, and syphilis, and prolonged exposure to X-rays. The fact that the US was apparently a party to a deal after the war, in which immunity was granted to the Japanese staff involved in exchange for details of the techniques developed (which included mass production of cholera and plague bacilli), can only reinforce apprehension concerning the integrity of some state-funded scientific research.

Studies in recent times have attempted to use scientific techniques to show that particular human groups or populations are innately more intelligent than others. Jensen (1969), an educational psychologist, concluded that the failure of certain 'compensatory education' programmes in the US was due to the innate inferiority of the groups concerned – mainly blacks. Eysenck (1971) supported Jensen's work, and argued that blacks, the Irish and the working class are relatively genetically inferior groups. IQ tests, they believe, indicated that a genetic meritocracy exists, and that variance between individuals cannot be accounted for by environmental factors. Barnes (1974) argued that Jensen's work is 'reasonably secure technically' and the 'indefeasibility of the 1969 paper is beyond dispute'. However, the idea that intelligence is mostly genetically based relies on the thesis that the number of intelligence genes is lower, on average, in certain groups of people; and as Rose (1976) points out, there is no such thing as a 'low IQ or a high IQ gene'; at best there may be particular combinations of genes which, in particular environments, produce 'high' or 'low' IQ. The relationship between 'genotypes' (DNA) and 'phenotypes' (the expression of the genes in the actual organism itself) is highly complex and not well understood. In Rose's view, the question of the contributions of genetics ('nature') and the environment ('nurture') to a particular trait like intelligence is not meaningful. That Jensen's work has encouraged the recommendation of programmes for sterilisation of 'inferior' groups is very disturbing.

Biotechnology poses weighty moral dilemmas for the scientists involved in this field, and has also created new threats to our ecology. Fears that a 'robotics revolution' using biochip technology may produce Frankenstein monsters may seem far-fetched, but the possibility that an alien gene placed in a common bacterium could create a seriously infectious new organism is very real. This problem may serve to limit developments in genetic engineering. As Kieffer (1979) observes:

when tampering with genetic control mechanisms, the danger of untoward side-effects is especially grave since one theory of cancer relates the disease to a genetic control system that has gone awry. The insertion of foreign

DNA may significantly increase the risk that the cell or tissue will inadvertently become oncogenic. For these reasons, this single problem of gene control could well prove to be unsolvable and may serve as the ultimate barrier to successful infusion of alien DNA into higher organisms.

Risks to the environment from these new developments in biology ('biohazards') include inadvertent changes in the biology of bacterial cells which could result in major epidemics, adverse affects on the human gene pool, and a possible irreversible attack on the biosphere which could threaten the evolutionary process. Another risk is that states may sieze upon the new technology in order to provide a 'genetic fix' for social problems. This combination of social engineering and genetic engineering is the sort of dystopian scenario often portrayed by science fiction writers.

5   Medicine in the Third World

In contrast to the overall health improvements in the developed countries, the Third World still experiences a high incidence of most diseases. The populations of these countries have not attained the required standards of nutrition and hygiene compatible with the improvements in health experienced by the developed countries over the last century, and cholera, plague, leprosy, typhoid and tuberculosis still thrive. Both infectious diseases and those associated with malnutrition remain prevalent, with over 50 per cent of all childhood mortality attributed to nutritional deficiencies. A poor diet lowers resistance to infectious diseases and considerably increases their severity once they have been contracted. Describing the situation in the Third World, Doyal (1981) writes:

Specific dietary deficiencies cause a great deal of chronic disease in under-developed countries. Lack of vitamin A causes many thousands of people in the Third World to go blind. In India, for example, some fifteen thousand children under five go blind each year as a result of this deficiency. Illnesses such as beri-beri, pellagra, and scurvy are also the result of particular dietary imbalances.

Western medicine and medical technology have little impact on health in the absence of adequate nutrition, and even exacerbate the health problems of the underdeveloped countries. Third World doctors trained in the West often attempt to run sophisticated hospital systems which are not only far too expensive to be sustained, but are in any event inappropriate to the health problems of these countries. In the Third World less than 15 per cent of the population have access to clean water supplies, at least 50 million people suffer from tuberculosis, 200 million suffer from malaria, and possibly one-quarter of the entire world population is thought to be infected by roundworms. Third World doctors who come to the West for post-graduate training often stay on, thus creating a 'brain-drain' from the Third World to the western countries.

Drugs account for a large portion (between 25 and 50 per cent) of the total health expenditure of underdeveloped countries. Multinational pharmaceutical companies hold over three-quarters of the drug patents in the Third World, and control most of the drug manufacture. Promotional activities often fail to warn the consumer of important side-effects, disastrous to the user's health. Other forms of technology transfer include hospital construction, and the provision of all the associated modern surgical and diagnostic equipment. Hospitals of this type distort the whole balance of Third World medical provision and tend only to benefit the wealthy élite of these countries.

The World Health Organisation (WHO), affiliated to the United Nations, has influenced medical policy in underdeveloped countries. Unfortunately, in past years the experts employed by WHO have tended to share the high technology orientation of western medicine. Although smallpox eradication has been achieved, their 'technical fix' response to Third World health problems has failed to eradicate other diseases, such as malaria, leprosy, cholera and typhoid. Population control measures have been encouraged as an integral part of aid programmes, but have failed to have any significant impact on population growth because the customs, beliefs and economic needs of the local communities involved have been ignored.

Sophisticated medicine and technology, for the most part are not paramount in containing the mortality and morbidity of the people in the Third World. It is suggested that holistic medicine with its consideration of the environment of the community, and the will of the people of the community, is the only way forward in the underdeveloped countries. The governments of India and some African countries are rejecting the western medical model and favouring the Chinese model, using 'bare-foot' doctors with sufficient knowledge to give basic medical care and perform an educative role in promoting preventative health practices and who live and work in rural communities. The Soviet Union too has traditionally offered an alternative, low-level medical personnel, the *feldshers,* who practise mainly in the rural districts. A network of these community doctors, who also usually have access to, and training from, city-based medical staff, is being instituted. Doyal (1981) warns, however, that as a result of this close association with professionals sharing the western model of medicine these rural medical schemes provide new mechanisms of social control over the peasants who are expected to put their own resources into health schemes over which they have no real control. Sadly, the health problems of Third World countries are likely to persist as long as the North–South gap is so great; and every year millions of people will continue to die from diseases which have long been understood and controlled in the West.

6   Health Economics and Freedom

Medical science and technology create ever increasing possibilities for new treatments. The Hippocratic oath imposes on doctors the duty to seek the best treatment available for their patients, but often this can be achieved only by increasing the costs of health provision. Not only is the direct cost of much modern medicine high, but the social costs of maintaining life can also be very considerable: life-saving surgery on the elderly, or on persons suffering from congenital diseases who often subsequently require continuous nursing care and medication, greatly increase the costs of health services to society. But, as Campbell (1978) observes, few doctors wish to abandon treatment purely because of the cost of life to society, though the allocation of limited medical resources often obliges them to do so. The costs of health services are borne either by the patient or by the general public, depending on the system of health provision a society operates. Britain and the US provide an interesting contrast: Britain has a nationalised health service, while the United States operates a health system in which about two-thirds is supported by private funds and institutions. In both systems costs of health care have risen sharply in recent years due to the increased availability of high-cost technology treatments, and the resulting problems in sustaining an up-to-date and efficient service have attracted the attention of some economists.

The difficulties of defining 'demand' for health services, 'efficiency of output', and 'net benefits' to society tend to render traditional economic analysis inadequate in informing social policies. This has not, however, deterred some economists from recommending a free-market system. It is suggested that such a system should, for example, replace the National Health Service in Britain (see Lees, 1961). This suggestion is based on the principle of consumer sovereignty ('individual preferences'), and is a direct appeal to individual freedom. Most economists feel more comfortable with the free-market model, the dominant paradigm in economic theory. The ideological objection to this model being applied to health services is that need and merit are divorced from economic demand, that is, the desire and ability to pay for services. The .limitations of 'ideal' competitive behaviour, where conditions of uncertainty prevail, render the description of reality supplied by the impersonal price system inadequate, because there are no markets for the bearing of some medical risks (Arrow, 1963). No company will wish to provide insurance for a person suffering from an incurable disease which is very expensive to treat.

Compared to most other industrial nations the USA has more physicians available for the population, the population averages more annual visits to the doctor, expenditure on health, measured both in total and in *per capita* terms, is the highest in the world, yet life-expectancy is lower, infant mortality higher, and the maldistribution of health resources greater there. The reasons for this failure to cope adequately with the health needs of its citizens are social as well as economic, but where emphasis is on consumer (patient) purchasing power, it is inevitable that those who most need medical care are least likely to get it. In the US gross inequalities of medical provision exist between rich and poor areas, and because surgery is more remunerative than other treatments, hospital surgeons find proportionately twice as many patients 'in need' of surgery as British surgeons (Campbell, 1978).

Not only does the demand side of the market equation for health services produce imbalances in resource allocation and inequities in health provision, but so does the supply side. It took the poisoning of some ninety people in 1937 in the US by a product 'elixir of Sulfanilamide', marketed by the Samuel E. Massengill Company, to persuade the American government that legislation on controlling the marketing of drug products was needed. The Federal Food, Drug and Cosmetic Act, 1938 resulted, and included in its provisions the stipulation that the manufacturer of any new drug had to present evidence of its safety. The producers were not, however, required to present evidence that the product was *effective,* and this led to the marketing of a number of medically useless compounds (Sjostrom and Nilsson, 1972).

The professional autonomy which the medical profession has in exercising controls over the form and extent of treatment virtually eliminates the freedom of choice of the individual. However, an oligopoly of firms exists, including La Roche, Distillers, Pfizer, Cyamid, Bristol and Upjohn, which control western manufacture, testing and marketing of drugs, and against this 'industrial-medical complex' the medical profession is virtually powerless. Needless to say, the individual consumer (patient) has even less power than the medical profession to exercise his or her autonomy against such a powerful commercial system.

The British government in 1980 failed in a legal action aimed at obtaining cost information on the most commonly prescribed tranquilliser Valium, when it first considered excessive profits were being made on this product by the Swiss firm La Roche. Also, it took many years for parents of 'thalidomide children' to obtain compensation from Distillers for the crippling damage caused by this teratogenic drug (see, for example, Sjostrom and Nilsson, 1972). In the opinion of Klass (1975), the drug industry creates confusion about drugs among doctors, avoids price competition, and because of its protective patent system its rate of new drug innovations is inadequate. He describes how, in the United States, private insurance schemes are controlled by ten powerful commercial insurance companies, the vested commercial interests of which block attempts to control democratically medical services, and re-distribute health resources.

Imbalances exist in all health systems throughout the world – the facilities for the mentally handicapped, subnormal and geriatric tend to be neglected in favour of the more prestigious areas of medicine, such as transplant surgery. Even in the Soviet Union there is an inequitable distribution of resources between urban and rural areas. Campbell (1978) argues that of all the health systems being operated in the developed world, the free enterprise approach of the US is the most inequitable. Indeed, the Federal government has recognised the inadequacy of the system and has introduced the centrally-funded Medicare and Medicaid programmes for the poor. But attempts to alter priorities are seen by those who gain most from the continuance of the free market in medicine as threats to commercial and professional freedom.

The fundamental issues in health care are ethical and social as well as technical, and thus demand the participation of society as a whole. Neither the medical profession nor the 'mandarins' of the drug industry are competent alone to make the sorts of decisions over health priorities, experimental research and human eugenics which confront modern society. Governments have recognised that professional self-regulation provides an inadequate safeguard to the consumer against the industrial–medical complex, and in most industrial nations state regulations have been brought into force. The freedom of scientific inquiry is constrained by codes of practice, and applies, for example, to experiments on humans, and for assessing at what point death has occurred. Experiments with recombinant DNA engendered the forming of the Genetic Manipulation Advisory Group (GMAG), now disbanded, in 1976 in the UK, and various ethics committees are now charged with the task of overseeing research and development. Professional Standards Review Organisations (PSROs) were introduced in 1972 by the US government to monitor the performance of doctors working within the federally-funded programme.

Holistic medicine, some have suggested, with its emphasis on preventive medicine could aid in providing the sort of environment, (including safer working conditions, better diet and the availability of unbiased health information) which would enhance good health and personal autonomy. The doctor–patient relationship could perhaps be transformed if health care became rooted in the communities in which people live and work, and the costs of health services both direct and indirect might be dramatically reduced. The moral and ethical decisions involved in abortion, psychiatric intervention, genetic engineering, euthanasia and the prolongation of life should be made by the community as a whole. Having membership of review committees drawn from a cross-section of the community, and not left by default to professional bodies and commercial interests, could help ensure the full participation of the community in these important areas of health.

References

Arrow, K. (1963), 'Uncertainty and the welfare economics of medical care', *American Economic Review*, Vol. LIII, 5, December, pp. 941-73.

Bajema, C.J. (1976), *Eugenics Then and Now* (Stroudsburg, Penn.: Dowden, Hutchinson & Ross).

Barnes, B. (1974), *Scientific Knowledge and Sociological Theory* (London: Routledge & Kegan Paul).

Baruch, G. and Treacher, A. (1978), *Psychiatry Observed* (London: Routledge & Kegan Paul).

Berke, J.H. (1979), *I Haven't Had to Go Mad Here* (New York: Dodd, Mead & Co.).

Campbell, A.V. (1978), *Medicine, Health and Justice* (New York: Longman).

Clare, A. (1976), *Psychiatry in Dissent: Controversial Issues in Thought and Practice* (London: Tavistock).

Darwin, C. (1859), *On The Origin of Species by Means of Natural Selection, or the Preservation of Favoured Races in the Struggle For Life* (London: John Murray; reprinted Harvard University Press, 1975).

Doyal, L. (1981), *The Political Economy of Health* (London: Pluto).

Eysenck, H. (1971), *Race, Intelligence and Education* (London: Temple Smith).

Friedson, E. (1975), *Profession of Medicine* (New York: Dodd, Mead & Co.).

Gillie, O. (1981), 'Judge quashed plea to broaden murder case', *Sunday Times*, 8 November.

Hawkins, D.F. and Elder, M.G. (1979), *Human Fertility Control – Theory and Practice* (London: Butterworths).

Illich, I. (1975), *Limits to Medicine* (Harmondsworth: Penguin).

Jensen, A.R. (1969), 'How much can we boost IQ and scholastic achievement?', *Harvard Educational Review*, 39, pp. 1-123.

Kennedy, I. (1981), *The Unmasking of Medicine* (London: Allen & Unwin).

Kieffer, G.H. (1979), *Bioethics: A Textbook of Issues* (London: Addison-Wesley).

Klass, A. (1975), *There's Gold in Them Thar Pills* (Harmondsworth: Penguin).

Lederman, E.K., (1970), *Philosophy and Medicine* (London: Tavistock).

Lees, D.S. (1961), *Health Through Choice: An Economic Study of the British National Health Service* (London: Institute of Economic Affairs).

Malthus, T. (1798), *Essay on the Principle of Population* (reprinted, London: Macmillan, 1909).

McKeown, T. (1979), *The Role of Medicine* (Oxford: Blackwell).

Mintz, M. (1969), *The Pill* (London: Hodder-Fawcett).

Morimura, S. and Shimozato, M. (1982), *The Devil's Gluttony* (Japan: Kappa).

Morris, N. (1960), 'Human relations in obstetric practice', *Lancet*, 23 April, pp. 913-15.

Norton, A. (1973), *Drugs, Science and Society* (Glasgow: Fontana).

Parsons, T. (1972), 'Definitions of health and illness in the light of American values and social structure', in Gartley Jaco, E. (ed.), *Patients, Physicians and Illness: A Source Book in Behavioural Science and Health* (New York: Free Press).

Pincus, G. (1975), *The Control of Fertility* (London: Academic Press).

Rose, S. (1976), 'Scientific racism and ideology: the IQ racket from Galton to Jensen', in Rose, H. and Rose, S., *The Political Economy of Science* (London: Macmillan).

Sjostrom, H. and Nilsson, R. (1972), *Thalidomide and Power of the Drug Companies* (Harmondsworth: Penguin).

Vessey, M.P. (1978), 'Contraceptive methods: risks and benefits', *British Medical Journal*, 9 September, pp. 721-2.

Yoxen, E. (1983), *The Gene Business* (London: Pan).

8 Energy

CHARLES BOYLE

1 Introduction

Without a large and steady supply of energy modern industrial society could not function. It is to science and technology that we owe the concept of energy, and the practical means of generating, distributing and using it in various suitable forms.

In this chapter we consider some important aspects of the energy question. Section 2 provides an historical background to our present consumption patterns, which are discussed in Section 4. Section 3 gives an account of important technical questions in the light of our current scientific theories. There is wide consensus, of course, about the basic principles of physics that relate to energy, but much less agreement on some of the technical problems associated with nuclear reactor safety or the disposal of radioactive waste. In Section 5 we consider some of the ways in which politics and economics enter into the energy field; and Section 6 looks towards the future provision of energy.

2 Historical Background

One of the most important scientific achievements of the nineteenth century was the formulation of the principle of conservation of energy. This great unifying principle brought together in a remarkable way not only the work of individual scientists of genius like Carnot, Joule, Kelvin, Liebig, Helmholtz, Mayer, Clausius and others, but also the hitherto disparate disciplines of mechanics, thermodynamics, chemistry, electricity and physiology. Scientists recognised that energy was an important factor in all the processes of the physical world.

It has not been till the second half of the twentieth century that there has been a wide public appreciation of the role that energy plays in economic and social processes too. A history of the availability and use of energy in the widest sense would be a history of mankind, involving food, material production, social organisation, environmental impact, and so on. The recent heightened public consciousness of energy use has occurred as some signs of future scarcity became evident, after a period of enormous increase in consumption during which large amounts of cheap energy were available – at least in the industrialised world.

When they discuss energy most scientists and technologists concentrate on questions of supply, which they tend to see in terms of technical problems, easily quantified and amenable to scientific analysis. It is at least as important, however, to bear in mind questions of demand, and also to appreciate the economic, political and social dimensions of the energy industries. Of the world's top ten multinational corporations, in terms of total sales in 1978, seven were oil companies and two were car companies, whose products are major energy consumers (*New Internationalist,* March 1980). Control over energy is a vital national and international political issue: it is mainly because of oil supply that the Middle East is one of the world's most politically sensitive regions. It is abundance of energy which underpins western consumer society and the life-style it generates. Energy, then, is of far more than mere technical significance.

Before the Industrial Revolution, human energy needs in the production of food and manufactured goods were met largely by direct heat from the sun, and by human and animal power supplemented to some slight extent by wind and water power: wood, dung and other materials were burned to provide heat, and still are in many parts of the Third World. By contrast, the nineteenth century, from an energy point of view, can be described as the age of coal; and the twentieth century has been increasingly the age of oil (Foley *et al.,* 1981). Coal played a key role in the Industrial Revolution. It was used in the smelting of iron for machinery, as a source of power in steam engines, and for domestic heating and cooking in the homes of a rapidly increasing population. The iron industry in Britain, hampered by lack of wood to make charcoal for smelting, was rejuvenated in the eighteenth century by the discovery that coal in the form of coke could be substituted. Large-scale coal-mining was made possible by the development of steam engines to pump

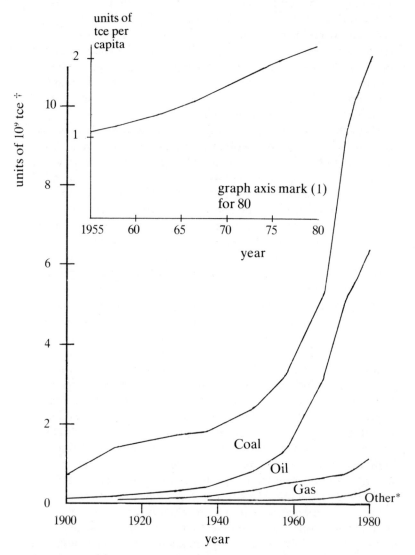

Figure 8.1: World trends in the consumption of commercial energy (1900–79)
Inset: World commercial energy consumption (tce per capita) (1955–79).
*'Other' includes nuclear, geothermal and hydroelectric power.
Source: UN *Statistical Yearbooks* and other UN publications.
†tonnes of coal equivalent (tce).

water from mines, and the coal produced was transported by water or rail for industrial and domestic use and for export. Total annual coal production in Britain increased from 10 to 60 million tonnes between 1800 and 1850, and then to 225 million tonnes in 1900. In Europe, too, industrialisation first occurred near coalfields and coal consumption soared. In the 1880s coal overtook wood as the primary energy source in the USA. (*National Geographic Magazine*, February 1981).

Though the production of coal world-wide is now annually some four times what it was in 1900, it has been displaced as the pre-eminent source of energy by oil (see Figure 8.1.). The oil industry originated in the USA, where much of the early technical expertise developed, and it has been largely dominated by American companies ever since, though recently a measure of power and influence has been wrested from them by the countries in whose territories the oil is found (Odell, 1981).

Of the seven major oil companies (the Seven Sisters as they are sometimes called), five are American (Exxon, Mobil, Texaco, Standard Oil California, Gulf), and of these five three had their origins in the splitting up in 1911, on the orders of the US Supreme Court, of John D. Rockefeller's Standard Oil, which held a near-monopoly, particularly in the refining industries, at the turn of the century (Sampson, 1975). (The other members of the Seven Sister family are Royal Dutch-Shell and BP.) All seven have interests not only in oil production but also in exploration, refining, petro-chemical production, shipping and marketing. Between the wars, either as individual companies or in joint ventures, they obtained cheap concessions all over the world, but particularly in the Middle East, and until the last decade or so they exercised such power that individual producing nations seeking more national control over their oil were usually forced to capitulate. Prices were fixed, more or less openly, by mutual agreement, and extremely rich profits were made.

First in America, and then in Europe, the boom in transport, particularly by private motor car, created an immense demand for petroleum products. In the US oil overtook coal as the leading energy source in the early 1950s, and the growing substitution of oil for coal was one of the consequences of increased American influence in Europe after the Second World War. Because of its relative cheapness to produce and the ease with which it can be transported, stored and used, oil remains today the leading fuel source, accounting for some 45 per cent of the world's commercial energy supplies.

The use of gas as a source of light has a history stretching back to early in the nineteenth century, when coal gas was first produced. The gas industry in Britain is of interest as it was one of the first modern industries in which government felt forced to intervene, setting up standards of purity and safety to allay public anxiety about the new energy technology. It is natural gas, of course, rather than coal gas which is important today. It has progressed from being considered an embarrassment, a useless by-product of oil wells shamefully wasted by being flared or allowed to blow away, to being seen as a most valuable fuel and a feedstock for the chemical industries, providing about a fifth of the world's total traded energy.

Electric power stations date from the 1880s, and demand for energy in the form of electricity has grown very rapidly since then, approximately doubling in each decade recently. This is not surprising since electricity is extremely convenient, and has a multitude of uses from lighting to powering electronic circuitry; it can equally well drive tiny or huge motors for a wide range of domestic and industrial purposes. The development of all these applications, the greatly increased efficiency of electric power generation, and the decrease in transmission losses are the result of enormous scientific and technical advances. Electricity is in some basic way associated with all the high technology of the twentieth century.

In the richer countries electrification has been an essential component of modern industrial development, but it requires a very high capital investment; and in the Third World it is a matter for debate just how much emphasis should be given to it in comparison with other pressing needs. The electricity supply industry is now government-controlled in most countries, and because of its size and its key economic role exerts a very powerful influence on overall energy policies.

Much the largest part of the world's electricity is generated in power stations fired by fossil fuels, but about a quarter comes from water power, and many hydro-electric schemes are in operation all over the world: all Norway's electricity is derived from water power.

Only a few per cent of the world total comes from nuclear fission. The earliest nuclear reactors were small research models, built during the Second World War in connection with the Manhattan Project in the USA. Later, larger reactors were constructed, especially for producing plutonium for nuclear bombs, but it was not until the mid-1950s that reactors came into operation which were specifically designed to deliver power, first to submarines, and then on a much larger scale to national grids. The military origins of nuclear power, the continuing link with nuclear weapons, and the secrecy surrounding the industry, have been factors in the failure to dispel widespread public distrust and suspicion (Patterson, 1977).

Other ways of harnessing energy have been historically important, but in quantitative terms at present are insignificant. Windmills, for example, date in Europe from the Middle Ages, and from even earlier times in China; they were to be counted in tens of thousands in Europe in the nineteenth century. Watermills, too, were important in rural economies though their power output by modern standards was very small.

The energy for agriculture in pre-industrial societies was supplied by animals and human beings. Animals have been replaced by machines and fertilisers, both requiring huge amounts of energy. Most workers have moved to towns, leaving only 5 per cent, or less, of the total workforce now working on farms in industrialised countries. It is thus in energy use for food production that some of the most spectacular differences between primitive and developed societies may be seen. Output (food) energy to input energy ratios have changed from values in the range 20 to 40 to much less than 1 (Leach, 1976), though on the other hand much less energy is wasted in cooking in modern kitchens than over open wood fires.

One final point worth mentioning in this brief account of the history of energy supply refers to changes in the distribution of consumption throughout the world. The fraction of the total energy that is consumed by the poor countries is steadily increasing, due to their growing consumption *per capita* and their growing populations; and though the USA (with only 6 per cent of the world's population) was consuming in 1974 some 30 per cent of the world's energy produced that year, this represents a considerable decline from the 47 per cent figure of 1947.

3   Technical Factors

Before going on to look at present patterns of national and international energy use, we outline a few of the important scientific and technical considerations which must underline or limit any energy proposals. The two basic principles of physics concerning energy are the first and second laws of thermodynamics – the first being the law of conservation of energy; the second, in one of its forms relating to heat engines, spelling out the theoretical limitations in extracting mechanical work from heat.

The first law deals with *quantities* of energy, and allows us to use one set of units for energy in all its inter-convertible forms – chemical, mechanical, electrical, nuclear, heat. In practice, of course, there is a luxurious profusion of units in the different energy industries. To most people it is surprising how much mechanical energy, as measured on any scale based on human exertions, is locked up in a small amount of fuel. The heat energy available in 1 kilogram of coal is about 8 kilowatt-hours, theoretically equivalent to the mechanical work which could be done by some hundred people working hard for one hour. We are right to think of modern energy supplies as providing each individual with a large number of 'energy slaves'.

For simplicity we shall mainly use as our energy units tonnes of coal equivalent (tce), one tce being approximately 8000 kilowatt-hours, and 0.6 tonnes of oil equivalent. Most scientists use joules for energy, and watts (joules per second) as units of power, the suffix (e) implying 'in the form of electricity' and (t) meaning 'thermal, in the form of heat or combustible fuels'. Oil men prefer to talk in terms of barrels of oil, and energy policy experts in the US, when discussing the energy consumption of nations, find 'quads' convenient, a quad being a quadrillon (10^{15}) BTUs or about 40 million tce. The situation is further complicated by the use of different conversion factors (from coal to oil) by different organisations such as the UN, the UK government and BP, not to mention short, long and metric tons, and coals of different types with different calorific values. A lack of standardisation of units of length, capacity and weight is common throughout the world, but in units of energy there is an especial diversity, though this is more of an irritant than a serious problem.

The second law of thermodynamics highlights the different

quality of different energy sources. For the purposes of obtaining electricity or useful mechanical work, it is much better to have a small mass of material at a very high temperature, rather than a large mass at a temperature only a few degrees above that of the environment, though equal *quantities* of heat might ultimately be given out in both cases during cooling. High temperature heat is high-grade. For maximum efficiencies in power stations operating temperatures should be high, but there are limits to the temperatures than even specially constructed boilers can withstand. For this reason the largest efficiency of conversion that can be achieved is about 35 per cent. This means that at best nearly two-thirds of the heat obtained by burning the primary fuel is lost. Indeed, it has been estimated (Thomas, 1977) that if we take into account transmission losses, and also all the energy used in the construction and assembly of the components of the power station, the efficiency is only about a quarter. Losses in end-use push the figure down even lower. Some of the heat at present lost in generating electricity can be used for district heating. Such combined heat and power (CHP) schemes, which are in operation in several countries but not to any great extent in the UK, depend on specially-designed power stations being located near the built-up areas they are to serve. Over half the total energy consumed in industrialised countries is used for space heating, and to use electricity directly for this purpose is extremely wasteful.

Many meteorologists are worried about the effects on the world's climate of the release of large quantities of heat and of carbon dioxide from the burning of fossil fuels, the so-called 'greenhouse effect', though they differ in their predictions of both the size and the direction (increase or decrease) of possible changes of temperature. One fear is that relatively small temperature variations could trigger quite large climatic effects, leading, for example, to expanding deserts and consequent threats of famine. Such concern provides a counsel of caution in the elaboration of high energy scenarios of the future involving ambitious fossil fuel programmes.

Many of the technical problems associated with energy centre on questions of conversion and storage which are crucial to the basic technical goals of the supply industries – the delivery of energy of the right types in the right quantities at the right time and in the right place. These technical aims, it is worth emphasising again, cannot be isolated from their economic, social and political contexts. What the right type and quantities of energy might be is a question involving factors such as price, social priorities, and political expediency, about which there may well be wide differences of opinion within any society. A confrontation of scientific and engineering problems, and an understanding of them, are necessary but not sufficient conditions for successful energy planning (*Scientific American*, 1979).

One basic problem is that of storage of electricity, or rather, since electrical energy is consumed as it is generated, it is the problem of converting electrical energy into some other form and then back again, when needed, into electricity without too much loss. National demand for electricity fluctuates from hour to hour, and from season to season. Having the capacity to supply power at the peak hour of the coldest day of the year means that some plants will be idle at periods of low demand; there is thus almost always a surplus generating capacity, even allowing for power stations which are out of service for overhaul or repairs. One solution (though an expensive one) is provided by pumped storage systems. Using electric power, water is pumped to higher levels during off-peak periods and allowed to flow back, generating electricity, later when it is needed. Other suggestions being investigated include the storage of energy underground in compressed air in caverns, or in huge rotating flywheels. Much research has been done into the conversion of electrical to chemical energy, but systems have yet to be found with low enough losses and using small enough quantities of cheap materials to make them economically viable for more than a few specialised purposes.

By considering nuclear power we can illustrate the technical difficulties that arise when modern high technology is being developed. Some of these are specific to the nuclear industry, some apply more generally. Foremost are questions of safety and danger, both to the workers in the industry and to the public at large, and it is to be noted that even in technical matters, especially when they concern new developments, such as huge increases in scale or more extreme operating conditions, there is by no means unanimity among the experts (Patterson, 1977).

With regard to reactor design it may be asked, for example, if the pressure vessels are strong enough – resistant enough to corrosion or crack formation and propagation; if power densities are so high as to cause a threat of melt-down in case of an accident; if emergency cooling systems are reliable; and so on. And in the event of things going wrong, what quantities of radioactivity would be released to the environment. What are the dangers of leaks in the system? The reactor is only one link in the chain of the nuclear fuel-cycle, which also includes mining and processing, reprocessing of spent fuel, transport of radioactive material and disposal of waste. Doubts have been expressed about each of these stages, but perhaps the most serious fears have been in connection with the as yet unsolved problem of waste disposal. Low-level radioactive wastes may be dispersed widely, but the long-term effects of the small increases in background radiation thus caused are by no means fully understood. High-level wastes are also accumulating. These substances, which remain dangerous for very many years, are at present kept from over heating, and are closely supervised in the hope that future generations will develop the technology to cope with them. Though it has always been common practice for us to leave behind political and economic muddles, this is perhaps the first example of an important demand on the technical ingenuity of our descendants. Some of those hostile to nuclear power argue that such demands are morally indefensible (Royal Commission, 1976, Council for Science and Society, 1979).

Advocates of 'high energy futures', and hence necessarily of high technology, pin much faith on fast-breeder reactors, and ultimately on power stations using nuclear fusion, which could be fuelled by virtually unlimited supplies of hydrogen isotopes from the water of the oceans. Fast-breeder reactors, which use violently reactive liquid sodium as a coolant, raise many of the same problems as conventional nuclear reactors, but in an even more acute form, while the harnessing of fusion energy in commercially exploitable quantities still seems remote. In the history of fusion research, promising paths ahead have frequently turned out to be blind-alleys.

A disadvantage of nuclear power stations is that they are large and preferably operated under steady output conditions near maximum capacity; they are not easy to switch on or off, or to turn up or down, and thus they add inflexibility to the distribution system. With wind and direct solar energy, on the other hand, one of the problems is the intermittent nature of the power source, raising once again the question of storage. Solar energy can of course be trapped on a large scale, but it can also be utilised in a multitude of small independent heating systems and electric generators. Flexible, small-scale, 'soft' technologies present technical problems no less challenging than those of 'hard' technology, but usually of quite a different nature. In the former case they are problems of diffuseness, in the latter of concentration. Conservation of energy is best achieved by a revolution in the public awareness of waste, but technical expertise is needed too in improving efficiencies of energy use in heating, in transport and in industry. The effort put into research on renewable energy sources and conservation is, however, very small in comparison with that expended on nuclear research. In the UK, for example, in the period 1975–80, £16.7 million was spent by the Department of Energy on research and development on new energy sources and conservation, while £665.3 million was spent on nuclear research (Cook, 1981).

We have made little reference so far to fossil fuels, and it would be wrong to give the impression that the technical dimension is any less important for them than for other energy sources. On the contrary, there are questions of safety and environmental degradation similar to those raised by nuclear power development; one has only to think of the dangers of super-tankers at sea and of coal-mining as an occupation, or of atmospheric and water pollution and the disfigurement of the countryside by slag heaps, to mention only a few examples. Research continues into improved or completely new techniques of recovery of larger fractions of fuel from mines and wells, into methods of extraction of oil from shales and tar sands, into the transport of fuels, into technologies of conversion from coal to oil or gas.

There is one further group of mainly, if not purely, technical problems that is extremely important and is frequently encountered in energy literature. With an eye on planning for the future we need to make as accurate estimates as possible of the world's reserves of the various types of non-renewable energy resources and the fractions of these reserves we might profitably recover at different price levels. This involves not only the exploration of the less well-known regions of the world and geophysical prospecting using a variety of methods, but also informed guesses about future advances in extraction technology. Because of all the uncertainties involved, sets of figures of proven and possible reserves show wide variations, reflecting the assumptions and prejudices of their authors (Freeman and Jahoda, 1979).

4 Present Energy Use

World energy consumption, which has increased to some twelve times the estimated 1900 figure, climbed till 1979 (see Figure 8.1), but has since declined. The industrialised nations have been making efforts to improve their conservation practices and to reduce their dependence on oil. As a result of these measures, and of the slowing-down of economic growth, energy consumption went down in some of these countries during the 1970s: for the UK, for example, 1973 was the peak year.

The average commercial energy consumed *per capita* is now some 2 tonnes of coal equivalent per year, but this global figure conceals a very wide variation between countries. There is a broad general correlation between a country's wealth and its energy consumption. Countries with very low gross national products *per capita* tend to consume very little energy *per capita*; these Third World countries lie very near the origin in Figure 8.2. On the other hand, it is by no means true that the wealthiest countries are the most profligate with energy. Switzerland can generate a GNP *per capita* about a quarter higher than that of the US using only one third of the US energy *per capita*. Canada, the UK, East Germany and the USSR, as well as the US, are big energy consumers in relation to their GNPs compared to the Scandinavian countries and Japan. There are many reasons why the ratio of energy to GNP should vary from country to country, and special factors come into operation in certain cases, but it is obviously of great importance in estimating future world energy demand to note that some rich industrial economies can be sustained with energy supplies *per capita* much lower than the current US figures. There is a big difference between a world policy that aims to supply 6 tce or less per inhabitant annually, and one that requires 12.

Total energy figures tell only part of the story; energy has to be delivered in the right form for its final uses – a car cannot run on coal, for instance, nor an electrical tool on oil. What mix of primary energy sources does a modern industrialised nation use? And where, in general, does the energy go? Tables 8.1 and 8.2 provide answers for the UK and the US. In Table 8.1 we note the large losses

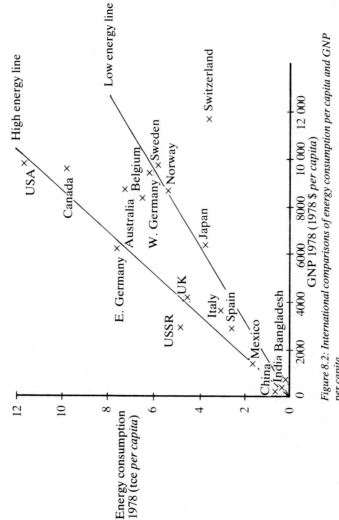

Figure 8.2: International comparisons of energy consumption per capita and GNP per capita

Source: The World Bank, *World Development Report*, 1980.

Table 8.1: Primary and secondary fuel consumption in the UK and the US, 1980 (units million tonnes of coal equivalent)

Primary fuels			Secondary fuels	
natural gas	UK 70 UK 830	power stations, refineries, coke ovens etc.	natural gas	UK 67 US 640
coal	UK 121 US 640		coal and coke products	UK 30 US 160
electricity*	UK 15 US 310		electricity	UK 32 US 370
crude petroleum	UK 121 US 1350	losses in conversion distribution and own use	petroleum products	UK 100 US 1240
total	UK 328 US 3130	UK 99 US 720	total	UK 229 US 2410

Sources: UK Department of Energy, *Energy Flow Chart, 1980.*
US DOE Report, *Securing America's Energy Future: The National Energy Policy Plan,* July 1981.

*This is hydro-electricity (UK 2; US 200) and nuclear electricity (UK 13; US 110). The US hydro figures include 'renewables'.

Fuel (UK)	Iron and steel industry	Other industry	Trans-port	Domestic	Other	Total
gas	2	23	0	34	8	67
coal products	6	9	0	13	2	30
electricity	1	10	1	12	9	32
petr. products	3	24	56	5	12	100
Total	12	66	57	64	30	229

Fuel (USA)	Industrial	Transport	Residential	Commercial	Total
gas	340	0	210	90	640
coal products	130	0	20	10	160
electricity	180	20	100	70	370
petr. products	300	720	100	120	1240
Total	950	740	430	290	2410
End-use losses	270	650	130	200	1250
Useful work	680	90	300	90	1160

Table 8.2: Final energy uses for UK and US, 1980, with US end-use losses
Units: tonnes of coal equivalent (in tce)

Sources: as for Table 8.1. In these tables the UK figures are rounded to the nearest unit; the US figures to the nearest ten.

(already referred to in the last section) incurred in power stations in the conversion mainly of coal and petroleum to electricity, and in the oil refineries which provide petroleum products and in particular gasoline or petrol. Table 8.2 gives details of the further losses incurred in end-uses in the US; we see that only 1160 units of useful work emerge from an input of 3130 energy units.

Substantial amounts of coal are used as a source of the coke which is required in the smelting of iron. Of the energy in the secondary fuels, industry consumes over a third, and transport 25-30 per cent (almost all in the form of petroleum products for cars and lorries). The domestic sector is also important. About a half of the primary energy supply, it is estimated, is used for space heating, mostly in private homes but also in commercial and public buildings and factories. Detailed lists of end-uses in various sectors and the associated consumption figures are important not only in accounting for the energy we produce, but also in pinpointing where savings can be made and higher efficiencies achieved.

5 Political and Economic Factors

Politics, taken to include economic and social considerations, comes into energy questions in many different ways and at different levels – international, national and local. We attempt here to outline a few general aspects of the political dimension, and then, mainly by way of illustration, to look in a little more detail at one type of energy that arouses strong emotions – nuclear energy.

5.1 General Considerations

Energy resources are not, like air, equally distributed over the globe; in their different forms they are much more strongly concentrated in some places than others, by geological or geographical determinants. Also, as we have seen, there is wide inequality between different countries as regards their levels of industrialisation, energy demands and technological capacities. These two facts lead, at the international level, to intense political manoeuvring.

Most discussions of international energy production begin with a recognition of the extraordinary situation of the Persian Gulf countries, and in particular Saudi Arabia. Because of their rich oil fields and their huge reserves, and because they consume very little, these countries are by far the greatest exporters of petroleum, and the US, Western Europe and Japan are very dependent on them. The US, in spite of being the world's largest coal and uranium producer, and the third largest oil producer (see Table 8.3) still needs such large extra oil supplies that it is the world's biggest oil importer. Western Europe has far smaller energy resources than the US – though it comes near to self-sufficiency in coal, it still needs to import oil. Within Western Europe, some countries, like the UK, are relatively well-off in coal or oil or both, but most others are not, and are major importers. The same is largely true of Eastern Europe. Japan is very poor in resources, and though it uses energy efficiently, its industries' demands make it the world's second largest oil importer and it needs to import 90 per cent of its energy. The USSR is a major producer of coal, oil and uranium, and also has vast reserves, relatively undeveloped, of natural gas. At present

Table 8.3: Energy production 1979

Coal (million tonnes)		Crude oil (million tonnes)		Petroleum products[b] (million tonnes)	
World	2718	World	3115	World	2420
1. USA	666	1. USSR	586	1. USA	688
2. China[a]	635	2. Saudi Arabia	476	2. Japan	213
3. USSR	495	3. USA	420	3. France	111

Natural Gas (billion m.^{3})		Uranium (000s tonnes) 1978	
World	1540	World	56.7
1. USA	552	1. USA	12.6
2. USSR	407	2. Canada	9.4
3. Netherlands	93	3. USSR	7.0

Main OECD Importers of Crude Oil
(million tonnes) 1979

1. USA	324	the major supplier
2. Japan	233	in each case was
3. France	126	Saudi Arabia

Source: 'The World in Figures', The Economist, 1980.

[a]including lignite
[b]excluding USSR

it is a net exporter of energy, especially to Eastern Europe. China is also relatively well endowed. The energy outlook is perhaps bleakest for those Third World countries, the great majority, which have few resources of their own apart from wood and waste materials, and tend to rely on expensive imported petroleum products as a source of energy for their fledgling industries (Ion, 1980).

At the national and international levels the politics of energy is extremely complex. It involves nations and governments anxious to ensure, because of its economic and strategic importance, continuous, guaranteed supplies of energy in appropriate forms, preferably from a variety of sources to avoid too great reliance on a particular one. It involves huge corporations like the international oil majors, the Seven Sisters already referred to, whose main concerns are with profits both in the short and long term. These companies, too, hedge their bets by operating in many countries, and in many spheres of activity – exploration, production, refining, distributing and marketing. They diversify by buying extensive interests in energy industries other than oil, (e.g. coal and nuclear power); they support intermittently research and development in new technologies such as the extraction of oil from shale and tar sands, and the harnessing of solar power. These companies have sales for exceeding the GNPs of many of the world's smaller nations, and can exert great economic pressures.

In addition, within countries and wholly or partly under state control, there is a whole series of powerful groups with (often conflicting) interests in energy, in Britain, for example, the Central Electricity Generating Board, the National Coal Board, British Gas, the Atomic Energy Authority, and so on. The trade unions are, in comparison, extremely weak financially, but are capable at times of mobilising support effectively. All these various organisations act through pressure groups and form shifting alliances; their interactions make energy studies difficult and intricate, far more than a simple review of technical problems and possibilities.

The international oil industry provides many examples of political and economic struggles where the stakes are high. Since the 1930s the smaller oil-producing nations have been engaged in continual battles to wrest control of their oil away from the international majors whose trump cards are their technical expertise in production and refining, and their market outlets. Mexico nationalised oil production in 1938 but found itself cut off from world markets. Governments in Venezuela in 1948 and in Iran in the early 1950s who challenged the companies' supremacy found themselves violently overthrown, and replaced by administrations servile, or at least better attuned to, the interests of the Seven Sisters, who in 1968 were still responsible for 78 per cent of world production, 61 per cent of world refining, and 56 per cent of world marketing facilities. By the late 1960s, however, their dominance was being eroded by smaller American companies who, because of import quotas imposed by the US in 1959, had been turning their attention increasingly outwards, and also by European groups such as the Italian state company ENI, who, in their operations with producing countries, were prepared to accept smaller shares of the profits. Producer countries also set up their own companies (Odell, 1981).

In 1973 the oil-producing and exporting countries (OPEC), adopting for the first time an effective united front, were able to hoist oil prices enormously, but it would be a mistake to think that the international majors came out of it badly. The profit made by Aramco, for example, on a barrel of crude oil, price $2.40 in January 1973, was 79c; this profit went up to $3.73 in January 1974 when the price was $10.83. (Aramco is the consortium of Exxon, Texaco, Socal and Mobil operating in Saudi Arabia.) There would seem to have been complicity between OPEC and the majors, but the blame for the price rise was placed squarely on OPEC. There were other reasons too, perhaps, why the change was not altogether unwelcome to the Americans. The price rise had a much worse impact on America's chief industrial competitors like Japan than on the US itself (Kaldor, 1979).

Japan and many Western European countries are keen to free themselves progressively from reliance on the majors, whose talk of 'rational' arrangements for oil production and refining is seen as thinly disguised self-interest. Direct deals with oil-producing countries are becoming increasingly common; these have the welcome side-effect of opening up further trade between the countries concerned. American anxiety about the construction of a pipeline to carry natural gas from Siberia to Western Europe led to attempts by the US to slow down or block the transfer of the technology needed for this project.

Finally, mention must be made of one important group which receives disproportionately little attention, in spite of the numbers that belong to it. This is the group of energy consumers. It has already been suggested that in the general consciousness of energy, questions of supply dominate over those of demand. This reflects the power of the suppliers compared to that of the consumers. The former, whether private or state-controlled, are large organisations, few in number, with many interests in common, rich, and extremely well-organised; the latter are numerous, very diverse in their interests, individually relatively poor, and disorganised. Many consumer groups are highly critical of the supply industries and feel that not nearly enough is done to analyse and cater for what they perceive, often in a wide environmental context, as the real energy needs of society.

5.2   Nuclear Energy

Among the most hostile and most articulate of these citizens' groups are the various anti-nuclear organisations, which may also embrace other bodies such as coal-mining trade unions whose interests are threatened by the expansion of nuclear power programmes. Though the nuclear contribution to the world's total energy production is still very small, the issues involved generate great controversy. Governments have tottered because of them; and they are discussed briefly here as a further illustration of the interaction of social, political and economic forces with technology. (Elliott *et al.*, 1978).

In an earlier section we have looked at some of the technical problems, especially those associated with safety. Safety can never be purely a technical matter since moral judgements about acceptable levels of risk are involved, even if we assume (and many would deny that we can) that risks can be assessed and quantified by experts who reach unanimous conclusions. Because very dangerous substances are produced, transported and disposed of, many social and political questions arise (Counter Information Services, 1978).

There are fears that a large nuclear industry undermines civil liberties. For example, it has been argued that workers in this industry should not have the same rights to take industrial action as other workers in view of the great potential dangers involved in, say, leaving high-level radioactive wastes unattended with no cooling system operating. The nuclear fuel cycle (see Figure 8.3) is vulnerable to terrorist attack at many different points, and to the theft or unlicensed transfer of nuclear weapons materials, perhaps in small quantities over long periods of time (disturbingly large amounts of 'material unaccounted for' have been reported). Police forces with special powers are thought to be necessary for protecting power stations and other plants, and for investigating employees in the nuclear and ancillary industries, their families and friends. With a greatly expanded programme for nuclear power in any industrialised state, many people would be involved. Is this sort of large-scale police activity acceptable in a 'free' democratic society? It is argued on one side that this is not a great price to pay for abundant supplies of electricity, on the other that the 'nuclear state' is inevitably a police state which even far greater material benefits could never justify.

The question giving rise to most concern, however, is perhaps

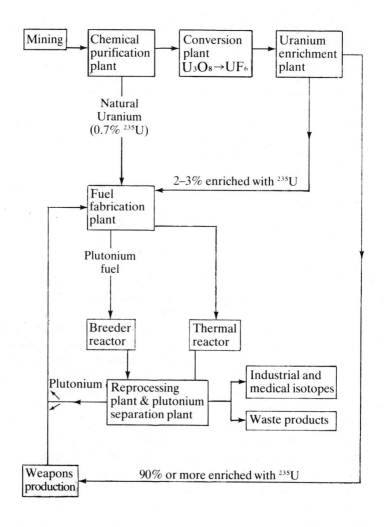

Figure 8.3: The nuclear fuel cycle
Source: SIPRI, 1975.

that of the military links with civil nuclear power and the horizontal proliferation of nuclear weapons, as more and more nations acquire nuclear technology. Any nuclear reactor can, in theory, be operated on a military or a civil fuel cycle. In the former the fuel is removed after a few months of operation and reprocessed to yield weapons-grade plutonium 239; in the latter it remains much longer to maximise energy production. (The early British reactor at Calder Hall was operated to produce significant quantities of both plutonium and electricity.) Countries which export nuclear reactors lay down internationally agreed safeguards to be observed by client nations, in order to minimise the possibilities for producing or diverting plutonium for military purposes. The International Atomic Energy Agency administers this system of safeguards, but it is far from foolproof as the growing number of 'nuclear-weapons powers' demonstrates. The more reactors there are operating, and the more shipments of fuel made for reprocessing, the easier it is becoming to divert materials for weapons production without detection.

The prospect of growing weapons stockpiles in dozens of nations led to the joint drafting by the USA and the USSR of the Non-Proliferation Treaty in 1968, signed by many – but by no means all – nations. Notable non-signatories include some half dozen nations who either already have, or are well advanced in, the construction of nuclear bombs. Such nations argue that they are perfectly justified in their actions as long as the super-powers fail to honour their undertaking substantially to reduce their own stockpiles.

Passing now to economic considerations, we might at first sight think that here is a clear-cut field where figures can be simply assigned and added up, and direct comparisons made between the costs of a unit of electricity from nuclear and coal- or oil-fired power stations. On closer inspection, however, and in the light of past experience, complicating factors emerge. The economics of nuclear power is just as controversial as any of its other aspects (*Ecologist*, 1981).

Designing and building power stations can take up to ten years or more, and there can be considerable delays both in starting up and completing construction, with huge associated expenditure. Over this period high and varying inflation and interest rates make for complications in accounting and for disagreements about 'real' costs. Improvements in technology occur, but some reactors perform much worse than expected. Increasingly stringent environ-mental and safety regulations lengthen construction times and add to costs. There are problems about the relative efficiencies to be assigned to power stations run on base-load and to those brought in only at peak periods. There are uncertainties about fuel costs. There is the question of separating the costs of civil and military programmes where these overlap. Opponents of nuclear power argue that not enough account is taken of the costs of storing high-level wastes for indefinite periods, of dismantling obsolete power stations, of policing the industry, and of insuring against accidents. They complain, too, in the UK, that they are refused access to information that would allow them to check official figures. In short, it is impossible to give a simple answer to the question: Is nuclear electricity cheaper and, if so, by how much? It all depends on your basic assumptions and accounting methods.

Certainly, there has been a great shift from the attitudes of the 1950s when it was stated, in a famous phrase, that nuclear electricity would be 'too cheap to meter'. The British nuclear programme over the last thirty years has been anything but a success in terms of returns on public investment. The industry has had to undergo several major restructurings; it has failed to export any reactors since 1959. It has been plagued by indecision over which reactors to develop to follow its early relatively successful Magnox ones, but the Advanced Gas-Cooled Reactors (AGRs) it has favoured have performed very badly, so that now there is a movement away from them towards the American-designed Pressurised Water Reactors (PWRs). The US programme is also beset with difficulties: apart from the Three Mile Island accident, we can take as an example the suspension of work in 1981 on two of the five proposed nuclear power stations in Washington State, where estimated costs escalated from under $4 billion to over $100 billion, and huge debts have been incurred, (*Sunday Times*, 22 November 1981). Orders world-wide for nuclear reactors reached a peak shortly after the 1973 oil crisis, but since then they have declined sharply and can-cellations have steeply increased. France, however, with its present reliance on energy imports, continues to press ahead with an ambitious nuclear programme.

It is illuminating to look at energy forecasts from the past, and to note how they have changed. The UK Electricity Council's short-range forecasts (six years ahead) of electricity demand were too low in the immediate post-war period from 1945 to 1950, but far too high during the 1960s and 1970s and have been considerably revised downwards recently. The existence of spare capacity weakens the case of those who advocate massive nuclear programmes. In the early 1970s in the USA forecasts by government agencies of installed nuclear capacity for the year 2000 were in the range 1200-1500 GWe. By 1980 these had declined to the range 150-200, figures that in 1970 would have been dismissed as absurd. (The actual installed nuclear capacity in 1979 was about 49 GWe. *Ecologist*, 1981.)

The debate on the economic viability of nuclear power continues. One side emphasises the need for huge new sources of energy if present industrial society is to continue to function and to develop in the Third World, given that the extraction of metals and other materials and the production of food will be increasingly energy-intensive, and that fossil fuels are being rapidly depleted. The other side emphasises the need for low energy use and strict conservation policies to protect the environment and to provide far more employment than an economy based on the less labour-intensive high technology industries can provide. There is a wide gap between the extreme positions. That there is still much to be learned about the problems can be illustrated by quoting from a report of the UK House of Commons Select Committee on Energy in 1981 that:

we were dismayed to find that, seven years after the first major oil price increases, the Department of Energy has no clear idea of whether investing around £1300 million in a single nuclear plant (or a smaller but still important amount in a fossil fuel system) is as cost-effective as spending a similar sum to promote energy conservation (Cook, 1981).

6 Energy Futures

Is there really an energy crisis? Are we going to run short of energy in the foreseeable future? One approach to these questions, the 'house-keeping approach', suggests that all we have to do is to work out our present rates of energy consumption, try to predict how they will change in the future, and then set these figures against the world's total energy reserves – both the recoverable fixed reserves locked in the earth, and the 'renewable' energy recoverable annually from the sun. This method suffers from obvious weaknesses: most of its figures are extremely uncertain, it lumps different kinds of energy together indiscriminately, it deals in world totals, failing to allow for regional differences in both production and consumption; but at least it gives us order-of-magnitude answers, and a very rough insight into some of the issues.

One estimate based on a wide survey of the literature is that, in units of 10^9tce, there are proven and possible reserves of fossil fuels of 1500-1600 units, and ultimately recoverable resources of 7500-13,000 units. If energy from nuclear fission is included, these figures go up to 1600-1700 and 17,000-24,000 respectively. Geothermal sources could make a further contribution. Renewable sources could provide 8 units per year using proved technology, and 10-70 units annually might well be possible eventually. World consumption at present, as we have seen, is about 9 units annually, about 2 tonnes of coal equivalent *per capita* for a world population of 4.5 billion. Even if the population doubled and the average consumption *per capita* per year trebled, this would give us a total annual consumption of about 54 units (Freeman and Jahoda, 1979).

The figures above give no cause for immediate alarm; we have decades to work out suitable world policies. They also demonstrate that many very different energy scenarios for the future can be envisaged. It might be possible for all our needs to be satisfied by developing renewable resources and making big efforts at conservation. On the other hand, we can imagine a high-energy scenario with individuals consuming on average at higher rates than present-day Americans, and all-out development of nuclear and other potential (Lovins, 1975, 1977; Leach *et al.*, 1979).

The figures do not, of course, resolve questions of high-versus low-energy futures. These problems, as we have seen in the last section, are not simply technical, but involve conflicting value-systems and social philosophies, which we shall examine further in Part III. One other parameter of great importance in estimating future world consumption is the rate at which the economic gap between rich and poor nations is closing – if indeed it is closing. This economic gap is also an energy gap. For a more equal world the *per capita* consumption of the Third World with its huge, rapidly increasing population will need to increase sharply, and this will push up world consumption greatly, even if that of the rich countries remains steady. It is argued that a low-energy scenario with high investment in the exploitation of energy from the sun is most likely to benefit the Third World. Those who favour greater equality fear that on present trends the Seven Sisters and the governments of the industrialised nations are biased strongly against developments such as these and are cutting back on them. But the energy situation is a fluid one. In a politically unstable world much can happen in the space of a few years.

References

Chapman, P. (1979), *Fuel's Paradise: Energy Options for Britain* (Harmondsworth: Penguin).

Cook, R. (1981), *No Nukes!* (London: Fabian Society).

Council for Science and Society (1979), *Deciding about Energy Policy* (London: CSS).

Counter Information Services (1978), *The Nuclear Disaster* (London: CIS).

The Ecologist (1981), vol. 11, no. 6, 'Nuclear energy: the real cost'.

Elliott, D. *et al.* (1978), *The Politics of Nuclear Power* (London: Pluto).

Foley, G. *et al.* (1981), *The Energy Question* (Harmondsworth: Penguin).

Freeman, C. and Jahoda, M. (eds.) (1979), *World Futures* (Oxford: Martin Robertson), Ch. 5.

Ion, D.C. (1980), *Availability of World Energy Resources* (London: Graham & Trotman).

Kaldor, M. (1979), *The Disintegrating West* (Harmondsworth: Penguin), ch. 6.

Leach, G. (1976), *Energy and Food Production* (Guildford: IPC Science and Technology Press).

Leach, G. *et al.* (1979), *A Low Energy Strategy for the UK* (London: International Institute for Environment and Development).

Lovins, A.B. (1975), *Non-Nuclear Futures: The Case for an Ethical Energy Strategy* (London: Friends of the Earth).

Lovins, A.B. (1977), *Soft Energy Paths* (Harmondsworth: Penguin).

Odell, P. (1981), *Oil and World Power* (Harmondsworth: Penguin).

Patterson, W.C. (1977), *Nuclear Power* (Harmondsworth:

Penguin).

Royal Commission on Environmental Pollution, 6th Report (1976), (The Flowers Report) Cmnd 6618 (London: HMSO).

Sampson, A. (1975), *The Seven Sisters* (London: Hodder & Stoughton).

Scientific American (1979), *Energy* (San Francisco: Freeman).

Thomas, J.A.G. (ed.) (1977), *Energy Analysis* (Guildford: IPC Science and Technology Press).

9 Military Technology

CHARLES BOYLE

1 Introduction

Leanings towards both hope and pessimism can be distinguished among the various opinions expressed about the ultimate origins of war. One optimistic view considers war as arising mainly because of material scarcity, resulting in fierce competition for limited resources of land, food or wealth. It follows from this view, which was held by some of the natural philosophers of the seventeenth century, that if science and technology can eliminate scarcity and attack disease, providing comfort and plenty for all, then discord both within and between nations will evaporate, and war become obsolete. Some pessimists, on the other hand, see violence as innate in man. Freud, for example, in his book *Civilization and its Discontents,* suggests that behind the veneer of civilisation there is a death-wish in all individuals, which periodically leads to the release of savage destructive forces.

These and similar questions have been given great urgency by the development in the present century of weapons of war capable of devastation and death on an unprecedented scale. An all-out nuclear war between the super-powers might well, under certain circumstances, wipe out most of the human race. It becomes imperative, if we are to survive, that we learn to control, by one means or another, our destructive potential.

Science and technology have played the key role in the development of this potential. It has been argued that by its very approach science is destructive, because of the emphasis in science (and in the whole Judeo-Christian intellectual tradition) on the control, domination and exploitation of nature, rather than on symbiotic cooperation with the natural world. Some indeed have seen this 'aggressive' scientific approach as essentially 'masculine', and have advocated a 'feminine' attitude to nature, more passive, respectful and ecological. It should not be surprising, it is said, that science has developed destructive methods so successfully. Be that as it may, what is incontrovertible is that without science and technology our modern weapons would not exist, and even pacifist scientists and apologists for science cannot wish them away.

The military applications of science and technology form no mere fringe activity indulged in by a few atypical tinkerers and experimentalists. In financial and manpower terms, they are central to modern research and development. As we argue below, in most of the large, technically-advanced nations, there is no other single field which consumes a higher proportion of the total research and development expenditure. Large numbers of scientists and technologists find employment directly in the military research sector; many others have contributed indirectly (and sometimes unknowingly) to it.

No study of the interactions of science, technology and society which leaves out the military field can be anything but unbalanced; yet among many scientists there is a tendency to ignore it, and a discernible reluctance to consider its relevance. It gives science a bad image, a bad name; it is much pleasanter to focus on pure research or research in agriculture or medicine, and regard this as more typical. These areas, however, attract sums of money that are only small fractions of those poured into research for defence purposes.

On a more positive note, it may be argued that wars, and the threat of wars, have stimulated much work that has had important civil applications. Military research has produced 'spin-offs', some of them perhaps doubtful blessings not unallayed with drawbacks arising from their military origins, and presenting other difficulties. They are bought, of course, at prices very much higher than would have been paid in purely civil programmes. Nuclear power stations owe much to the work in the early 1940s on the atomic bomb. Radar and similar detection and navigation systems that date from roughly the same period have important applications in the safe operation of shipping and aircraft services. The whole development of computers, transistors, integrated circuits and microelectronics was given a great boost by military needs for compact and reliable communications, guidance and monitoring systems. Missiles and satellites can be exploited for their peaceful as well as their prestigious and aggressive possibilities. Drugs such as penicillin, and pesticides like DDT, used on a large scale originally in the control of diseases in armies in the Second World War, are equally applicable to the diseases of civilians. The Germans in the First World War, cut off from their supplies of nitrates for explosives, invented a process for the fixation of atmospheric nitrogen which can be used efficaciously to produce agricultural fertilisers. A detailed list of such militarily-inspired innovations would be a very long one, and only a few major examples are mentioned here.

In the rest of this chapter we sketch out, first, a brief historical background to the developments in modern military technology which are reviewed in Section 3; while Section 4 looks at the resulting enormous drain on world resources, and at other economic factors. In Section 5 we conclude with a brief speculation on possible military futures.

2 Historical Background

Ingenuity applied to maiming, killing and weapon-making is to be found in most cultures stretching back to antiquity. Many respected figures (for example, Archimedes, Leonardo da Vinci, Galileo, Descartes) applied technological innovations to the war efforts of the ages in which they lived. The 'engineer' was originally a designer of 'engines' in the sense of 'machines of war', but these were essentially one-off individual contributions. It was the great general social changes in organisation and production brought about by the Scientific and Industrial Revolutions which really transformed war.

J.K. Galbraith chooses a revealing example from the military field as an illustration of the operation of what he calls 'the technological imperative'. He asks us to consider the differences between the Spanish Armada – in its day the greatest fleet the world had seen – and the modern American Navy. The latter is the product of huge capital sums and large numbers of trained specialists working over many (mainly peace-time) years on highly coordinated projects; it draws on many of the different resources of a highly industrialised economy. The Spanish Armada, by contrast, was built and assembled in a few months with little detailed planning (Galbraith, 1972).

War has become 'total', and scientific. It has ceased to be a matter of scratch armies locked in hand-to-hand struggle on a rural battlefield, where each individual saw at close quarters the injuries he inflicted. It has been, in the twentieth century, much more a question of whole, competing systems of production, with all their resources mobilised, and with civilian populations of men, women and children almost as involved and almost as much at risk as the professional armed forces who now, typically, wound and kill at very long range. War is not just killing the enemy, nor even simply, as Clausewitz said, the continuation of politics by other means; it is also a cultural activity with its own ritual, and its built-in restraints, without which the human race would long ago have been extinct. Modern technology has undermined some of these restraints.

Various aspects of the industrialisation of war were apparent during the American Civil War, 1861–5. Mass armies were deployed involving more than 4 million soldiers in all. Railways became important in mobilising them and in keeping them supplied with munitions and provisions. Many new types of military equipment were tried out – torpedoes, mines, grenades, machine guns, balloons, iron ships, submarines – but it was the rifle which was the most significant, giving as it did the advantage to defenders in field entrenchments rather than to attackers trying to over-run them (Fuller, 1972).

Artillery and rifles were steadily improved until the First World War, which continued the trends of the American Civil War on a much larger scale. But the generals, unable to learn from even the most horrific experience, and still loath to recognise that the bullet, the spade and barbed wire loaded the odds heavily against offensives, time and again pushed millions to death for gains of a few pitted and sodden miles of territory. There were 1 million German, French and British casualties in the Battle of the Somme between July and November 1916 alone; and altogether some 10 million perished in the war which was for long periods one of attrition.

Two technological developments, however, used effectively only fairly late in the war, tilted the balance away from fixed entrenchment and towards mobility: these were tanks, and poison gases, at first chlorine, and then later mustard gas and phosgene in shells (Clarke, 1969). Tanks, with submarines and aeroplanes, were also to play a very important role in the Second World War, though poison gas was not resorted to. Aeroplanes were widely used in the systematic bombing of civilian targets, particularly in Germany, though this was inhumane, wasteful and militarily ineffective. (The atomic bombs dropped on Japan are a special case.)

Technology has had a very important influence on twentieth-century wars in another sense: through the ubiquitous systems of internal and international communications it has provided, the widespread dissemination of propaganda has been possible. Also, at the ideological level, even 'democratic' modern states at war feel obliged to increase their control over their citizens by revoking many civil rights and liberties, such as freedom of speech, and by exerting a tight grip over the economy.

On the subject of war, what can we learn from history about our present situation? The old advice, 'If you want peace, prepare for war', seems not to have been particularly successful in preventing wars except for short periods: arms races have usually ended in war. And yet

at the height of the rearmament period immediately before the First World War, some 3-3½ per cent of total world output was given over to military uses, (while) by 1968 the share had risen above 7 per cent; the amount of resources devoted to military purposes in that year exceeded the total world output of 1900 (Freeman and Jahoda, 1978).

Even the most famous arms races of the past were puny by comparison with those of today.

Appalled by a stream of increasingly lethal innovations, writers on various occasions have considered the weapons technology of their time so destructive as effectively to outlaw war, making it unprofitable to both sides, if not totally irrational. This view was widely argued in the 1930s, and had been expounded at the turn of the century by Bloch, a Jewish banker, who in his attempts to promote peace forecast the course of the First World War with uncanny accuracy (Fuller, 1972). Yet such warnings were to no avail. It is not reassuring that the view twenty or thirty years ago of nuclear conflict as unthinkable seems now to be giving way to ideas of winnable limited nuclear wars.

3 Modern Developments

3.1 Nuclear Weapons
At the human interest level, the complex story of the development

of nuclear weapons is full of episodes that are deeply ironic. It was a man of strong pacifist inclinations, Einstein, who, as the best-known scientist in the world at the time, signed a letter to President Roosevelt that was influential in the early stages. The brilliantly successful Oppenheimer, leader of the scientists in the Manhattan Project to build the atomic bomb in the USA, was later to fall into disgrace, and to be refused access to nuclear secrets. Sakharov, one of the architects of the Soviet hydrogen bomb, was to end up as an outspoken and persecuted leader of the Russian dissidents. Scientists, many of them, by another twist of irony, Jews expelled by Hitler from Europe, reluctantly joined the Manhattan Project because of their fears of Hitler developing the bomb first, only to see it used, instead, without warning, against the Japanese, though it was known in military circles quite soon after the Allied invasion of Europe, that Germany, concentrating her technological efforts on rocketry, was far from having the capability to develop a nuclear weapon. This information was not, however, passed on to the Manhattan Project scientists. Finally we might mention that after the war an impoverished Britain, under a blanket of great secrecy, devoted huge resources to the production of its own bomb, only to find it largely irrelevant in its efforts to maintain itself as a leading world power (Jungk, 1960; Gowing, 1974; Freedman, 1980).

There is space here to mention only a few key dates: the discovery of nuclear fission by Hahn and Strassman in Berlin in 1938, the realisation shortly afterwards of the feasibility of a nuclear chain reaction, the explosion of bombs at Hiroshima and Nagasaki in 1945, the first Russian test of a fission bomb in 1949, the first American test of a hydrogen bomb in 1954, followed by a similar Russian test in 1955. The arms race was under way (Table 9.1).

Table 9.1: Total number (to nearest 10) of nuclear warheads, excluding bombers

Year	1961	1965	1970	1975	1979	1985
USA	160	1400	2900	6830	7590	8300
USSR	—	260	1740	4710	6740	?

Source: Pentz, 1980

By and large, the Americans took the initiatives with the Russians responding, though the launch of Sputnik in 1957 gave the latter at least a temporary lead in the development of intercontinental ballistic missiles (ICBMs). At the political rather than the technological level, the North Atlantic Treaty Organisation (NATO) was formed in 1949, and the Warsaw Pact Organisation (WPO) in 1955. One of the main concerns of the NATO alliance has been to prevent splitting effects due to differences of emphasis in policy and differences of interests between the US and the European members. Western European countries have been concerned to ensure that in the event of an attack on them by the numerically, though not necessarily also militarily, stronger conventional armed forces of Eastern Europe, the US would not abandon them. But a reluctance to build up very expensive conventional forces led to the idea of a nuclear response to a Russian attack.

As nuclear weapons stockpiles began to grow the concept of MAD – Mutually Assured Destruction – developed. Many believe that this doctrine of deterrence has been a key factor in preventing war in Europe between East and West; others argue that the reasons why war has not broken out are very different. Deterrence is based on the destructive power of nuclear weapons. Five main effects are usually listed in the writings of technical experts – blast, direct nuclear radiation, direct thermal radiation, the electromagnetic pulse and radioactive fall-out. The relative importance of these effects in inflicting injury and death depends on factors such as the design and explosive power of the bomb, the altitude at which it is exploded, and weather conditions. A few simple figures may be given by way of illustration.

A 1 Mt (i.e. megaton, 1 million tons TNT equivalent) air-burst at 8000 feet could be expected to destroy light buildings and severely damage heavier ones over a radius of 4½ miles, killing 50 per cent and injuring 40 per cent of the people in this area, due to blast effects alone. The explosion of a similar bomb at ground level would cause a crater 1000 feet in diameter and 200 feet deep. The earth and other material from such a crater would be sucked into the 'mushroom cloud', and much of it, highly radioactive, would be deposited over an area extending some hundreds of miles downwind, and several miles across, causing many deaths over a period of years. Worse fall-out could be obtained by bombing nuclear power stations. Thermal radiation effects (intense light and heat), preceding the blast by a few seconds could, in clear weather, produce severe burns and fires over hundreds of square miles, and might give rise to fire-storms that consume everything in their path. Direct nuclear radiation is of significance only in the case of smaller weapons or specially designed enhanced-radiation (neutron) bombs, where blast is minimised. The electro-magnetic pulse is important in so far as it causes high voltage surges in electrical equipment, thus disrupting communications.

Any estimate of the combined or so-called synergistic effects must take into account, as well as the factors listed above, shock and decreased resistance to disease in survivors, lack of hospitals, sanitation, food and so on. The Hiroshima bomb of 12,500 tons TNT equivalent, in an air-burst causing almost no radioactive fall-out, killed 68,000 people according to official figures, though many more died of injuries over the next five years. Some Japanese scientists have suggested 140,000 deaths in all. The US Congress Office of Technology Assessment (1980) estimates that Detroit, with a population of 4.3 millions, could be almost totally destroyed by a 25 Mt air-burst, with 1.8 millions killed and 1.4 millions injured; and in a total exchange of 7800 Mt between the US and USSR, the number of immediate deaths is estimated as 265 millions with 133 millions injured.

Both the US and the USSR have three main types of carriers for their strategic nuclear weapons: intercontinental ballistic missiles (ICBMs), kept below ground in 'hardened' silos; submarine-launched ballistic missiles (SLBMs); and bomber forces. These constitute the 'strategic triads' of the two sides. Estimates of their strengths are given in table 9.2 from which it can be seen that there are also 20-30 thousand tactical nuclear weapons, with an average explosive power ten or twenty times that of the Hiroshima bomb. Such tactical weapons, of which perhaps 10 thousand are in Europe, are designed for battlefield use, but in the densely populated European continent the collateral damage to towns and cities would necessarily be very great, however accurately they were directed at military targets. The UK and France have smaller stockpiles of strategic and tactical weapons, and in recent years other nations have joined the 'nuclear club'.

The doctrine of deterrence proposes that war becomes impossible, since no counterforce first strike (that is, a pre-emptive surprise attack on weapons carriers) can ensure the elimination of enough enemy missiles to prevent a highly destructive counter-

Table 9.2: Nuclear arsenals (1980)

Nation	Strategic		Other systems		Total	
	Warheads	Yield (Mt)	Warheads	Yield (Mt)	Warheads	Yield (Mt)
USA	9000-11000	3000-4000	16000-22000	1000-4000	25000-33000	4000-8000
USSR	6000-7500	5000-8000	5000-8000	2000-3000	11000-15000	7000-11000
Total (including UK, France, China)					37000-50000	11000-22000

Source: SIPRI, 1981a.

attack. But the whole thrust of the technological developments of the last decade has been to undermine this doctrine. First, there has been the deployment of bombs in numbers far beyond those required for deterrence. This was partly achieved by the production by the US and then later by the USSR, of missiles with several warheads, initially not separately controllable, but now with the warheads fixed in multiple, independently-targetable re-entry vehicles (MIRVs). Secondly, great efforts have been made to increase the accuracy of missiles whether fired from land, sea or air, so that there are now missiles with a good chance of hitting a particular building or silo, rather than just a particular city. Thirdly, further destabilising effects have been caused by scientific advances in submarine detection. Originally submarines, lurking in unknown positions, were meant to guarantee a massive retaliatory counter-attack in the event of an enemy strike, but now they are becoming increasingly vulnerable. There has also been much work on the military exploitation of satellites in connection with the developments listed above, and also for other purposes (SIPRI, 1982). Satellite anti-ballistic missile systems may be the next twist of the screw.

The overall effect of these developments is to heighten tension, to make the enemy fear that it can no longer wait for a first strike and then assess the destruction caused before launching a considered counter-attack; rather it will be tempted to retaliate strongly on warning or on suspicion, or perhaps even to get a strike in first.

The possibilities of war through false warnings, computer malfunctions and other accidents are greatly enhanced: reports from the US of some of the accidents involving nuclear weapons which have already occurred do not inspire confidence. On the other hand, the emphasis on accuracy and the stockpiling of a wide range of tactical weapons seem to indicate that limited nuclear wars are envisaged, with flexible responses selected from a whole range of options. The Russians maintain they will reply to any nuclear attack with a massive nuclear counter-attack, but some Kremlinologists claim to have detected evidence of 'limited-war' thinking in the writings of Soviet theorists.

Part of the normal cold-war strategy is to try to make the enemy confused and uncertain. It is claimed, for example, by advocates of the 'independent' British deterrent that it is useful in inducing an element of doubt in the minds of the Russians. The danger is of course that, with subtle and sophisticated doctrines, there is also the possibility of uncertainty, confusion and misunderstanding in the minds of one's own ranks and in those of one's allies. It has been aptly said that much strategic thinking is enveloped in a sort of nuclear fog.

3.2   Other Developments

Although it is often nuclear weapons which, because of their awesome destructive power, attract most attention, the pervasive and systematic application of scientific and technological ingenuity is clearly apparent in many other military fields. Only a few well-known examples are mentioned here (SIPRI, 1982). Particularly indiscriminate and inhumane are the various anti-personnel bombs designed to fling out over a wide area large numbers of fragments which become embedded in the bodies of men, women, children and animals who happen to be in the vicinity of the explosion. These fragments may be in the form of small barbed arrows (flechettes), which are particularly difficult to extract from flesh without extensive tearing; or they may be made of plastic so that they cannot be detected by X-rays; they may even be contaminated by radio-active material. The military advantage of such weapons is that they wound severely, rather than kill quickly, and thus absorb considerable enemy time and resources in the care of the injured. Other weapons widely used in Vietnam, for example, were booby-traps and cleverly disguised mines, exploded after a time-lapse, or by remote control or when approached by a living creature. Incendiary weapons, incorporating substances like napalm, are also regarded by most people as specially barbaric and cruel. But attempts in UN conferences to ban them have met with only limited success.

Chemical and biological weapons form another group which is attracting increasing research and development, in spite of public revulsion against them since the First World War and the hitherto relatively effective Geneva Protocol of 1925, which outlaws their use. Nerve gases have been produced which are many times more toxic than the chemicals used in 1914-18. They do not need to be inhaled to be lethal – a droplet of less than 1 mg absorbed through the skin is sufficient – thus special suits covering the whole body, and not just gas masks, are required for protection. Civilian populations are again at greater risk from this technology than appropriately equipped soldiers. The most recent important innovation in this field is the binary weapon – a shell containing two relatively harmless substances separated by a thin wall which is perforated after the shell is fired, allowing the chemicals to interact and produce a deadly nerve gas. A very wide spectrum of chemical weapons is available; it includes not only the nerve gases of greater or less volatility, but also crowd-control agents like CS and various incapacitating gases, and defoliating agents such as those used on a vast scale in Vietnam to destroy forests, crops and vegetation.

The use of biological weapons in the past to spread disease or contaminate food and water was not unknown, though it has not been common. But the discoveries and the techniques of molecular biology, genetic engineering and biotechnology have opened up whole new horizons with hitherto unrecognised potentialities for manipulating the basic units and processes of living organisms. Most experts on disease have been, and are, bent on eradicating it, but there are small groups among them actively involved in promoting it, selectively but deliberately, as a new part of the armoury of total war.

Since the earliest days of radio, electronics has been closely linked to the military field. As we have already noted, the improved guidance, detection and communications systems it has made possible have been readily exploited for purposes of war, and military demands in turn have led to major innovations in the field of electronics. The far-fetched speculations of science fiction of half a century ago have become, or are rapidly becoming, realities, such has been the pace of development over the last decades. There are the elaborate command, control, communications and intelligence (C^3I) networks of the super-powers, encompassing listening-posts, computers and satellites, and carrying endless torrents of coded information from sources as different as tapped telephone lines and ocean surveillance cameras in space.

The last decade has also seen great improvements in precision guided munitions (PGM), using laser, radar and infra-red devices for homing in on targets such as tanks. Probabilities of hitting targets have increased dramatically, but electronic counter-measures (ECM) can be taken, which in turn can be minimised by electronic counter-counter-measures (ECCM), and so on. Costs soar when equipment incorporates advanced electronics, and, as the Vietnam War demonstrated, the highest technology is not always the most militarily effective. The extrapolation of some present trends, however, leads inexorably to the idea of future wars fought on electronic battlefields littered with sensors, between rival groups of computers and robots. Where human beings come into the picture, after pressing the first button, is not at all clear!

Finally, in our review of modern military developments, some mention must be made of guerilla wars and of counter-insurgency and internal security measures. Here, too, science and technology play an important role, strengthening the security forces and the police by providing them with new weapons and electronic means of surveillance and control. These may ultimately be of little avail where there is a corrupt regime in power and strong support for insurgents among the population at large. It would also seem that, in some ways, highly centralised, industrialised societies are vulnerable to sabotage attacks at key points by small, well-organised, armed groups, but to what extent only the future can tell (Ackroyd *et al.*, 1977).

4   Economic Factors

We present below some figures on military and other spending. These figures, like almost all those given in economic statistics are not to be regarded as highly accurate – to better than 20 per cent, say – and some are worse. They are subject to many uncertainties rising from estimates, guesses, extrapolation errors, difficulties in definitions and classifications, different approaches in different countries and from different sources, and so on. The natural scientist, when presenting an experimental result, indicates its accuracy either implicitly through the number of significant figures given, or explicitly by stating the estimated error. However, rough figures, even orders of magnitude, are much better than no figures at all.

It must also be noted that, for propaganda purposes, there is often deliberate distortion or falsification of military statistics, the tendency usually being to give underestimates of one's own military force, and over-estimates of the threat by an enemy. USSR statistics, being largely secret, are extremely uncertain. The figures which follow are taken from SIPRI, 1981 and Sivard, 1981 which claim an unbiased and objective view. These sources should be consulted for further detail.

World military expenditure in 1981 was some $500 billion (1 billion = 1 thousand million). In constant (1978) US dollars, between 1960 and 1980 in each five-year period it increased by about 55 billion from 240 to 460 billion annually. This represents a huge and expanding drain of resources in almost every country in the world. As advanced electronics and other sophisticated tech nologies are incorporated in weapons, unit prices rise very steeply. Tanks and aircraft that cost tens of thousands of dollars each some forty years ago now cost millions. Whether their military effectiveness in all circumstances has increased by the same factor is open to doubt, though their potential for destruction is of course greater. Increased military expenditure does not necessarily lead to increased security (Smith, 1980).

Figures are presented in Table 9.3 to indicate military and other spending as compared to gross national product (GNP), a nation's GNP being to some extent an index of its prosperity, although it excludes goods and services in the informal sector (e.g. in households). NATO's share of world military expenditure is some 43 per

Table 9.3: Comparative expenditures in 1978 in billions of (1978) US dollars

		World	US	USSR	UK	Developing countries
	GNP	9279	2133	970	312	2076
	Military	418	109	103	15	97
Public	Education	474	123	49	17	77
	Health	326	74	21	15	31

Source: Sivard, 1981

cent, compared to the Warsaw Pact Organisation's estimated 26 per cent.

Detailed analyses of the expenditure figures reveals a huge and very rapidly growing arms trade (some $25 billion annually), with the main exporters of major weapons (aircraft, missiles, armoured vehicles, warships) in the period 1977–80 being the US (43.3 per cent), the USSR (27.4 per cent), France (10.8 per cent), Italy (4.0 per cent) and the UK (3.7 per cent). The main importing regions were the Middle East (32 per cent), the industrialised countries (31 per cent) and the Far East (10.4 per cent) (SIPRI, 1981).

From the point of view of science and technology, it is the relative sums assigned to military and other research and development which are important. Table 9.4 gives allocations under three headings.

Table 9.4: Government spending on research and development, 1970-80 (billions of US dollars)

	Military	Space	Civilian
US	117	36	79
EEC	36	7	115

Source: Sivard, 1981

It can be seen that over half the government funds made available for research and development in the US go to the military sector. In the USSR spending is probably on much the same scale. The UK government also puts about half its research and development effort into military research, largely in the aerospace and electronics industries, but most other European governments favour a far higher proportion of civilian or economically-motivated research. Much of the research and development associated with space is also of military interest, and the armed forces' involvement in the US space programme has steadily increased. One result of this war-directed bias is that a considerable proportion of the world's best scientists are involved in military projects of one kind or another.

The point is often forcefully made that there is a negative correlation between productivity and economic growth on the one hand, and military expenditure and research and development on the other. Japan, the country with by far the highest annual rate of growth in manufacturing productivity during the last twenty years (over 8 per cent), is one of the smallest military spenders. Italy and West Germany, too, until recently spent relatively little on arms, and had very rapid economic growth. The worst economic performers, the US and the UK, are the heaviest military spenders. Some commentators regard the USSR as an even clearer example of a country whose commitment to military growth has seriously impeded it economically.

It is sometimes argued that high military expenditure produces economic benefits by providing jobs and training and by modernising pre-industrial Third World societies. The case for this is very weak. Civilian expenditure by governments can generate far more jobs for a given sum and develop skills in the labour force much more useful to society. Strong military influences in Third World countries have, with few exceptions, had disastrous economic, political and social consequences. All too frequently military *coups* (some 76 in the last twenty years) have brought to power dictators whose first acts are typically the suspension of civil liberties and the denial of other elementary human rights. Before long they have instigated widespread repression, torture, terror and violence against the luckless civilians who are bold enough to protest against them.

Malign social and ideological influences are apparent even in countries in which the armed forces do not directly exercise total political power. President Eisenhower, in a famous speech he made shortly before leaving the Presidency, warned of the dangerous power of the military–industrial complex in the US. There are close and profitable links between the military top brass and the leaders of the small number of extremely powerful multinational corporations, which control the lion's share of the work of supplying military hardware, from the initial research and development stages to the final manufactured product. Possibilities for corruption proliferate, and scandals surface from time to time. If distortions of the goals of universal, disinterested knowledge, freely available to all, arise from the industrialisation of science, how much worse they become in the effective militarisation of science and scientists! Is knowledge that is secret, classified and primarily geared to destructive ends, true science at all?

5   The Future

In Chapter 12 we describe the displacement of traditional and religious categories of thought by a modern scientific approach, all-embracing and totalitarian in its scope, mathematical, analytical and experimental in its methods, and with the ultimate aim of control over the physical and social environment. Such an approach to waging war is clearly evident in recent military developments. It makes anachronistic concepts such as valour, chivalry and military honour, and focuses instead on systematic technologies of destruction using all the resources of modern science from molecular biology to operational research, and incorporating experimental innovations of every conceivable kind. But the problem is that increasing destructive power does not give greater control over the enemy. Beyond a certain point it threatens, on a planet of limited size and with fragile life systems, to produce not simply defeat for one side or the other, but the destruction of both. Self-defence, paradoxically, becomes suicide. In the words of Captain Ahab in the novel *Moby Dick:* 'Our means are sane, only our ends are mad'.

What are the main possibilities in the future? Slow continuous changes with present trends maintained indefinitely, or sharp discontinuities, breaks with present practice? Even if, as at present, outbursts of mass violence can be contained except for limited wars in the Third World, the nuclear and other arms races cannot continue for ever, consuming more and more scarce resources. Many indices point towards a short, but exceedingly destructive, war that would devastate Europe, the USSR and the USA and would also have severe effects all over the world. It must be in the interests of all to strive to the utmost to prevent such a war.

The other, more hopeful, discontinuities have to do with disarmament, and there are two main schools of thought here. The first sees the intensification of present efforts in the UN and elsewhere as leading to a slowing-down and then a reversal of the arms build-up through gradual, general disarmament. Since the Second World War there have been a number of arms control agreements which, though very limited in scope, can serve as a basis for further discussions (SIPRI, 1982). The second school is impatient at the lack of success of present policies and negotiating methods, and points out that the 1970s, which saw military expenditures and weapons stockpiles grow to unparalleled heights, were meant to be, and were officially designated as, the Decade of Disarmament. The supporters of this line of thought argue that only radical and dramatic measures will break the vicious spiral of stimulus and escalating response that provides the framework for the arms procurement policies of the super-powers. Such moves would probably need to be initiated by civilian peace groups and others outside the military-industrial complexes of the power blocs (Thompson and Smith, 1981).

The various approaches to disarmament are linked to, or follow from, various theories of the arms race, each reflecting a view of the interactions of science, technology and society. There are those who find that the blame for the present situation lies with scientists and technologists who push the military into accepting new and refined weapons for which they have no real need. Some say that the determinants of the arms race are predominantly to do with international, bureaucratic or even electoral politics. Others point accusing fingers at military strategists. Yet other theories regard military expenditure as one of the chief ways in which the capitalist world disposes of the surplus value it creates; or they postulate other economic determinants. Illustrations can readily be found giving plausibility to each of the theories in turn – a reflection of the great complexity of the phenomenon, and the great difficulties to be faced in overcoming vested interests, in promoting the conversion of arms industries to the production of more socially-desirable goods and in achieving substantial disarmament.

The problem is urgent, for if the present nuclear powers fail to agree on arms reductions or at the very least further arms limitations, there will be little to stop the spread of nuclear weapons to many more countries in the next decade with consequent grave threats to stability. If arms reduction does begin in the near future, new and more modest defence policies will be necessary, perhaps with the emphasis genuinely on defence and on the repulsion of invaders. Resources could thus be liberated which could have an enormous impact on the problems the world faces. When we look at the various threats to world peace pessimism comes easily, but the situation is not yet one in which to abandon hope.

References

Ackroyd, C. *et al.* (1977), *The Technology of Political Control* (Harmondsworth: Penguin).

Clarke, R. (1969), *We All Fall Down: The Prospects of Chemical and Biological Warfare* (Harmondsworth: Penguin).

Freedman, L. (1980), *Britain and Nuclear Weapons* (London: Macmillan).

Freeman, C. and Jahoda, M. (eds.) (1978), *World Futures: The Great Debate* (Oxford: Martin Robertson), ch. 10.

Fuller, J.F.C. (1972), *The Conduct of War 1789-1961* (London: Eyre Methuen).

Galbraith, J.K. (1972), *The New Industrial State* (Harmondsworth: Penguin).

Gowing, M. (1974), *Independence and Deterrence* (2 vols) (London: Macmillan).

Jungk, R. (1960), *Brighter than a Thousand Suns* (Harmondsworth: Penguin).

Office of Technology Assessment, US Congress, (1980), *The Effects of Nuclear War,* (London: Croom Helm).

Pentz, M. (1980), *Towards the Final Abyss? The State of the Nuclear Arms Race* (London: J.D. Bernal Peace Library).

SIPRI (Stockholm International Peace Research Institute) (1981, 1982), *World Armaments and Disarmament: 1981 and 1982 Yearbooks* (London: Taylor & Francis).

SIPRI (1981a), *Nuclear Radiation in Warfare* (London: Taylor & Francis).

Sivard, R.L. (1981), *World Military and Social Expenditures* (Leesburg, Virginia: World Priorities Inc.).

Smith, D. (1980), *The Defence of the Realm in the 1980s* (London: Croom Helm).

Thompson, E.P. and Smith, D. (eds.) (1981), *Protest and Survive* (Harmondsworth: Penguin).

10 Telecommunications and the Mass Media

BRIAN STURGESS

1 Introduction

Technological advances in communications are set to make the 1980s and 1990s a challenging time. In the West, if these developments are to reach mass-market production and diffusion, the necessary driving force is likely to be the prospect of profitable investment. Much advanced communications technology is financially risky, especially when the product or system produced is likely to have a small number of buyers – this can severely constrain innovations. Also, the existence of legal monopolies, or regulated enterprises, can hamper technological development, as witnessed in the development of digital telephone systems. Government support for technological innovation (by, for example, taxation policies, direct government investment, regulation or deregulation) needs to be evaluated. Such policy-decisions are likely to have significant effects on jobs, work, leisure and the form of our communication systems and mass media.

Communication involves the exchange of ideas between people. It has been defined as social interaction through messages (see Gerbner, 1967). Communication media are carriers which transmit messages; and mass media are 'all the impersonal means of communication by which visual or auditory messages are transmitted directly to audiences' (Gould and Kalb, 1964). The information transmitted can take a variety of forms: it can be, for example, for diversion or amusement, or for commercial, administrative or military purposes. It is useful to consider the social and political questions relating to the nature and composition of a 'mass' audience, and the technical questions relating to the means by which messages are transmitted.

This chapter deals with the economic, social and political aspects of change arising from new communications technologies. In particular, we consider advances that have brought together the computer and the means of communication. One of the main points considered is that of the imposition of restrictions intended to extend political and social controls over access to the means of communication. Section 2 is a brief commentary on changes in the technology of communications through the ages. This is followed by a discussion of some important recent developments in communications technology and the press. The economics of these technological changes is considered in Section 4. Emphasis is placed upon the potentialities of cable television and direct broadcasting by satellite (DBS), and the effects of new technology on the press. The telecommunications industry is discussed in terms of the resources it employs, and the role of government intervention and control in this industry. Section 5 considers political aspects of telecommunications and the mass media, control of world communications and the need to ensure data privacy, and discusses public concern over increasing political surveillance. Finally, the chapter concludes with comments on the future potentialities of telecommunications technology.

2   Historical Developments

In earliest times transfers of information were realised only through the medium of face-to-face meetings employing language. Despite the importance of the evolution of language in the development of human society, the complexity of organisations that can be built upon word-of-mouth communication is severely limited. A modern industrial company or military machine would be, of course, impossible. Even simple organisations would embrace problems because of a phenomenon known as 'control loss'. If a message has to be transmitted from one person to another through a chain from the sender to its final recipient, slight differences in the perception of the message by each link can become magnified until, as the number of links through which it must pass increases, its meaning becomes distorted. The diversity of languages illustrates another problem which hindered early communication between people and limited the complexity and scope of early organisations, confining them to small areas or linguistically homogeneous groups. The Old Testament myth of the Tower of Babel is perhaps the earliest recorded story of communications failure in an organisation when the common language needed to coordinate building efforts was suddenly transformed into a multiplicity of tongues.

The development of human society has been linked with the development of means of recording, storing, transmitting and analysing the information that could at first only be communicated orally. An obvious method of hastening the speed of information exchange and increasing the distance over which it can be accurately transferred is to provide fast transport to convey messages or written documents. Until the predominance of the electronic media in the twentieth century, advances in communications technology had been closely linked to advances in transport. The most recent advances in communications weaken this link, with the technical feasibility of the video conference and telecommuting.

Another important change has been the enhancement of the recording of information. The inventions of paper and printing, both originating in China, were, until the twentieth century, the most important means of exchanging and recording information. Printing and paper were cost-reducing innovations which allowed a vast amount of information to be recorded, stored and distributed. Block printing seems to have been used in the late years of the T'ang dynasty in the tenth century AD, and printing with movable type was invented in China in 1040. However, literacy was concentrated in the élite in China, partly because of the difficult nature of the Chinese written language with over 4000 characters, but also because the élite never allowed the power of literacy to be widely disseminated. This illustrates an important argument of this chapter, that the impact of technology depends upon the political framework in which it is applied. Printing did not have the impact on Chinese society that characterised its introduction into European societies. Although many European rulers attempted to control printing, they had little success. Books, always an expensive luxury, became, if not widespread, then certainly available in large enough numbers to influence political consciousness in some sections of society. The availability of pamphlets and tracts fuelled the Reformation and the English Civil War, and also proved important during the American and French Revolutions. The age of persuasion and political propaganda had begun.

The history of the press as an independent new medium free from censorship began in Holland. The principle arrived in Britain in

1695, almost by default, when the Printing Act, which aimed at controlling the number of presses and apprentices, was not renewed, due more to a failure to agree on changed controls, than to a desire to free the press. In the eighteenth century the numbers and circulation of newspapers expanded, and technical changes soon led to an even greater expansion.

The development of railways made possible a rapid country-wide circulation of London-based newspapers, but the provincial press remained economically viable for a number of years. From 1865 continuous rolls of paper could be fed into the presses allowing 20,000 impressions per hour. By 1914, with the daily circulation of some papers in millions, and with the growth of popular journalism independent of patronage (but dependent on commercial success, and increasingly on advertising revenue) the press seemed to be *the* medium of mass communication. Yet, from small beginnings in the nineteenth century, electronic media started to expand and proved to have an even greater impact on the development of mass communication than the press.

As early as 1837, Morse, an American, had made a circuit of copper wire which was able to transmit coded messages carried by an electric current. In 1844 a telegraph line able to transmit instantaneous coded messages was laid between Baltimore and Washington. To ease communications, in 1843 the Paddington to Slough railway line was linked by telegraph, and within four years 4000 miles of telegraph lines had been laid in the United Kingdom. Money orders were first sent along wires in 1850, and in 1851 the stock exchanges of London and Paris were linked. In 1865 a cable was laid across the Atlantic, by 1873 direct contact was made between Europe and Australia. The telegraph companies were either state-controlled or very soon administered by the state, but an opportunity existed for the growth of private news agencies such as Reuters.

The next major development in the growth of electronic media was the invention of the telephone which was a significant breakthrough because messages did not need decoding. In 1876, the first message was transmitted, and by 1880 the US had 54,000 telephones, 2 million by 1900 and 8.7 million by 1912. In Europe the dissemination of the telephone was achieved most quickly in Germany; in Britain it was slowed down by the insistence of the Post Office that telephone communications should be a state monopoly.

Although the technology for multiple lines and group communication existed, this form of telecommunication development was constrained.

The invention of radio, television and the development of the cinema had profound affects on society. Wireless waves were discovered in the 1880s, but it was not until the establishment of the Wireless Telegraph Company by Marconi in 1897 that their use was exploited. After initial transmission of radio messages and their use in communicating between ships at sea, it took until the end of the First World War before radio broadcasts were used for entertainment, information and political purposes, taking advantage of the possibility of reaching directly into the homes of the masses. The development of the thermionic valve pioneered in commercial uses by AT & T improved radio transmission and reception possibilities.

In 1922, under government control, a broadcasting authority was set up in Britain; while in America broadcasting remained in private hands, being licensed by the authorities. By 1938, 26 million families in the US had radio sets. The employment of radio broadcasting for the dissemination of political propaganda was extensive in Soviet Russia and in Nazi Germany. The cinema, which developed rapidly in the 1920s and 1930s as big business in the area of entertainment in America, was also exploited by the totalitarian states. Goebbels, the Nazi propaganda minister, set up a Reichs Chamber of films, which was to develop the themes of racial superiority and the triumphal rise to power of the Nazi Party.

The cinema was displaced as a mass medium by the rise of television broadcasting. Baird transmitted the first television pictures in 1924, and began intermittent broadcasting until his company was bought by the BBC in 1931. Regular broadcasting began in England in 1936, but it was in the US that television developed most rapidly. By 1970, with the widespread introduction of colour, television had become the main form of entertainment in the developed world. Yet because of the 'spectator' nature of broadcasting, the possibility of indoctrination, the emphasis on violence common in many programmes, and the fusion of fiction and reality that can occur, this medium has led to many unresolved debates about its role, and with the multiplication of channels and the possibilities of video recording, television is likely to remain a most controversial medium well into the future.

3 Modern Developments in Telecommunications

Early applications of electronics to communications and information processing were limited because they relied on valves. The invention of the transistor in 1948 in the Bell Telephone Laboratories in the US was a landmark in modern solid-state electronics. A transistor is a slice of a solid material which acts as a semi-conductor, and which reproduces the switching functions of a valve more efficiently in terms of speed, size, reliability, power requirements and cost. Silicon is the most commonly used material. A 'silicon chip' is a flat piece of material upon which a number of transistors can be imprinted. In response to commercial and military pressures a spectacular growth in the switching capacity of chips has occurred, from about 10 transistors per 'chip' in 1960, to more than 30,000 in 1977, and to around 500,000 in 1980. The average cost of production per chip declines rapidly as output is increased. Most of the fixed costs involved are in the design and construction of the master chip. This chip can then be replicated almost without limit using silicon, a plentiful and cheap raw material.

The technological advances in the field of communications are linked to the growth of the information-processing sector in modern economies. This sector has expanded because complex organisations require fast, efficient means of communication. It is the advent of the micro-chip, with its many applications in the field of communications, which allow vast quantities of information to be generated, transmitted, stored, retrieved and analysed. For example in the printing and newspaper industries, word-processors allow computerised text-handling by storing the text and newspaper layout in magnetic memory form, which is easily alterable without the need of continual resetting. Micro-chip technology has enabled British Telecom to introduce 'System X' – a telephone system using a solid-state switching system integrated with computer control.

The other communications revolution relating to television which is promising far-reaching social and economic changes is the introduction of cable TV. Cable systems are not new, especially in the US, where the number of cable subscribers increased from 9 million in 1975 to 18 million by 1981. Canada, too, uses more cable communication than Europe, and more than 60 per cent of Canadian homes receive television by cable, a higher penetration figure than that for the US. In Columbus, Ohio, a two-way facility allows viewers to react to television programmes and even to vote in referenda. At present in Britain about 2.6 million households (14 per cent of those with TV sets) receive services by cable television. Some systems were set up in the 1930s to distribute radio broadcasts. Recently the government has licensed thirteen pilot schemes covering 110,000 homes to test the market.

Modern cable systems are based on coaxial cables with a capacity of up to 40-50 channels, and on optical fibres with almost unlimited capacity, which can almost immediately offer extra television entertainment channels for consumers, but in the future will be capable of delivering information, and financial or other services into the home. Links could be created between businesses and homes in a way that might drastically change shopping and working habits by the 1990s. Entertainment channels using cable retain the spectator nature of television as a medium, but the extra capacity of the cables opens up the possibility of allowing two-way interactive information exchanges.

Advances in micro-electronics and in rocketry have led to the establishment of a world-wide communications network linked by satellite. The launch of the world's first communications satellite, Telstar, was in 1962. Four years later with two satellites, one over the Pacific and one over the Indian Ocean, full global coverage was achieved. This system of global satellite communications is known as INTELSAT. Satellites have become big business for the designers, manufacturers and launchers. By 1981, over 200 earth stations had been set up, with thirteen satellites in orbit.

Competing modes for future launchings are the US space shuttle and the French Ariane rocket developed for the European Space Agency (ESA). British Aerospace is a main contractor in the project which aims to put Europe's largest communications satellite (L–Sat) into orbit by 1986. This satellite is for the European Communications System managed by EUTELSAT. There are also plans for a Franco-German satellite for Direct TV broadcasting (DBS), a UK DBS satellite (UNISAT) in 1986, and a commercial satellite from Luxembourg which will broadcast in several languages.

Most information is currently transmitted by either modulating the amplitude (AM) or the frequency (FM) of carrier waves. An alternative is pulse code modulation (PCM), which transmits

Table 10.1: Telecommunications industry structure, 1980

Country	Ownership	Per cent of total telephones that are government operated	Per cent of industry output purchased	Major suppliers	Imports as percentage of domestic market
Australia[1]	Australian Telecommunication Commission (ATC) state-owned	100%	75%	L.M. Ericsson Pty. Plessey Australia STC Pty. (ITT) Have over 70% of the market	24%
France	Direction Générale des Télécommunications (DGT) state-owned	100%	61%	CIT–Alcatol Thomson–CSF CGCT (ITT) Have over 20% of the market	4%
West Germany	**Deutsche Bundespost (DBP)** state-owned	100%	85%	Siemens SEL (ITT) DTK Tekade Highly concentrated	6.5%
Japan	Nippon Telephone and Telegraph NTT state-owned	96.1%	44%	Hitachi Nippon Electric Fujitsu OKI Have over 90% of the market	2.7%

Sweden	Swedish Telecommunications Administration (STA) state-owned	100%	NA	L.M. Ericsson Tell **(STA)** Highly concentrated	26%
United Kingdom	British Post Office (BPO) state-owned[2]	99.3%	67%	General Electric Company Plessey **Standard Telephones and Cables (ITT)** Pye-TMC Have 90% of the market	9%
United States	**ATT – have 82% of telephones; independents – have 18%** Both are private and government regulated	0.15%	NA	Western Electric (ATT) General Telephone and Electronics North Electric (ITT) Northern Telecom Stromberg-Carlson Four-firm concentration ratio is 90%	1.92%

Sources: For per cent of telephones that are government operated, see ATT Longlines, *The World's Telephones*, (New York: ATT Longlines, 1978).
For per cent of industry output purchased by a carrier and supply concentration ratios various sources were used.
For imports as a percentages of the domestic market, statistics were taken from United Nations, *1978 Yearbook of International Trade Statistics* (New York: UN, 1979), vol. 11.
Diodati, J. (1980), 'Vertical integration as a determinant of industry performance in the telecommunications industry: an international comparison', *EARIE Conference Paper* (Milan, 1980).
¹The Australian figures are for 1975.
²The British government plans to denationalise British Telecom in 1984.

information in the form of a code of numerical values which can then be converted into binary units for transmission. Electronic switching devices, and especially solid-state, digital methods in telephone systems, have the added advantage of eliminating much 'noise' gathered in normal transmissions. Furthermore, they allow powerful and fast computer systems to handle communications because they process the data in the same form as that required by computers. All information, spoken and written – books, catalogues, reports and papers – can be stored in binary units on magnetic tape, on discs, or in micro-chips having a memory function. Instant world-wide communication, beamed by satellite and accessed through computer terminals, has become possible. At present almost two-thirds of intercontinental calls linking the world's estimated 500 million telephones are conveyed by means of satellite.

The use of satellites has allowed world economies to be linked together with the transfer of billions of dollars within minutes. Such transactions would have been thought impossible a century ago.

The opportunities for commercial use and for personal access of information are enormous. The satellite communications network is also of vital importance in conveying military, political and diplomatic information across the globe. The exchange of information does not always lead to a result acceptable to all parties, particularly if the information communicated is interpreted differently by opposing sides. Although the satellite communications network provides an opportunity to increase international understanding it may also be used as an instrument to heighten conflict.

4   Economics of Telecommunications and the Mass Media

4.1   Telephones

In this industry, the state is considerably involved – as a buyer of communications network, as a monopoly producer, and in connection with the research, development and production of the equipment. Table 10.1 illustrates the extent of state involvement world-wide, where twelve companies sell 80 per cent of world equipment. Several reasons are usually given for this close state supervision and control, and for the degree of involvement between suppliers and users in this industry. Table 10.1 shows that the operator of the telephone network is in a monopoly position in most industrialised countries, except the US where the dominant firm, AT & T, has an 82 per cent market share, and the network is usually state-owned. Whatever the ownership arrangement, most suppliers of telecommunications equipment face a single buyer, and have close links vertically with the national system in a manner similar to the links between the suppliers of military hardware and the government as contractor in the US. Table 10.1 also shows the degree of 'horizontal' concentration in the supply industry, and the importance of imports as a potential competitive alternative source of supply for national buyers. The nature of the 'vertical' links between supplier and user, parallels 'horizontal' links between suppliers. The telecommunications industry is far removed from the idea of free competition. The dangers of inefficiency, higher costs and the abuses of monopoly power are thus present.

Economic reasons are argued for the structure of the telecommunications industry. The arguments for a single national body are based upon economies of scale: the average costs are thought to fall as the scale of the operations increases. It is argued that a particular service could always be provided more cheaply by one monopoly operator, rather than by, say, two companies, both operating at less than optimum scale. These economies are thought to be present in research and development expenditures also, and are compounded by the large, risky investments required to provide a national network. For example, it has been suggested that the money needed to develop an electronic switching system could be anything up to $500 million (at the time of writing) and would take at least ten years. This type of investment could only be carried out by very large concerns. The case for state monopoly is argued because of the available evidence on economies of scale, but it may still be that state monopolies have been in operation, in many cases, for political rather than economic reasons.

Advantages of closer links between suppliers and users of telephone equipment can lead to more efficient transactions if a number of conditions are fulfilled which would not be the case with pure market exchanges. For example, if the product to be exchanged is complex and there is some uncertainty involved in the transaction from future outcomes, if it is purchased infrequently, and if unique risky investments have to be made, then as Williamson (1976) suggests, vertical 'hierarchical trading' arrangements are often preferable to 'free market transactions'.

The national telecommunications systems of most countries exhibit these features in the development of central office 'electronic switching equipment'. Yet the unhappy history of the UK telecommunications industry, and the development of 'System X', illustrate the extra costs and inefficiencies that arise from replacing the market with close state involvement with a number of large corporations. The former British Post Office, part of which is now British Telecom, decided in the early 1950s to move straight from mechanical switching into electronic switching, and research and development was carried out jointly with the suppliers. After an initial failure in 1962 each supplier undertook its own research, abandoning the joint project. Development followed a piecemeal approach until the early 1970s, when a decision was made to pool research, and to introduce a unified system throughout the whole Post Office – this was 'System X', a solid-state digital switching system, completely integrated with computer control. The intention was to launch the UK into the world telecommunications market as a major exporter of the system, and to improve the quality of domestic service by applying the latest technology. By 1982, nine years after its announcement, 'System X' seemed doomed to failure, having failed to attract one export order. By the mid-1980s 'System X' will be available in only 150 out of 6000 exchanges in Britain. Lack of co-operation between the supplier companies, and the single-minded desire of British Telecom to develop a national system, have been blamed for the problems. As the Carter Report (1977) on the Post Office commented,

if the only purpose was to give the UK a better and cheaper telephone system quickly, it might well be a desirable solution to buy designs on license from abroad and adapt and manufacture them in the UK. To do this however would involve withdrawal from a prime place in the technology of telecommunications.

4.2   Cable Television

The current licensing and regulatory system in Britain is not suitable, without legal changes, to allow large-scale development of cable, yet for political reasons, and from the pressure of vested interests, despite its enthusiasm for cable, the present government seems unwilling to deregulate, and allow free competition for the provision of cable TV services. The urge to supervise and retain control may strangle the 'cable revolution', because the initial investment sums are likely to be large. It has been estimated, at the time of writing, that £2.5 billion would be needed to lay a network capable of reaching the 50 per cent of the UK population living in the large towns and cities, yet important employment-generating effects, and possible valuable exports, could arise out of this development (ITAP, 1982). Concern from potential investors and operating companies is expressed that excessive supervision and regulation will curtail initiatives. As we have seen in other chapters of this book, a conflict exists between the idea of regulating technological development centrally, especially in high technology, military and other fields, and that of allowing the market to perform most resource-allocating functions. Nor is this conflict easily resolved. The *Economist* (6 March, 1982), that clarion of deregulation and the free market, has stated its views clearly on this issue:

'In the new information era the way forward lies through competition and diversity, not through monopoly or centralised controls.'

Yet in the field of high technology industries, competition rarely meets the classical model of rivalry among many small-sized firms; rather it exhibits features of rivalry between a few, very powerful, industrial complexes. This is certainly true in the production of communications hardware: cable operation, as a result of franchises, may be placed in the position of a local monopoly and may require regulation to control abuses. The market, research and other commercial opportunities are large, and banks, credit companies and newspapers have been willing to invest large sums of money in cable schemes in the US with no prospect of early returns. The services which are piped into the home and which can be censored, or cut-off by the operator, can be financed on a subscription, or a pay-as-you-use or watch basis, with perhaps a fixed standing connection charge. In the US a decision of the Federal Communications Commission in 1975, which relaxed many of the regulations on cable operations, paved the way for a rapid expansion of these services. Amidst a background of high competition for cable franchises, the US experience is that the network or national operators face declining audiences in a situation of a multiplicity of channels with wider choices of viewing.

In Britain, to protect the two national broadcasters, it is envisaged that it would be mandatory for cable operators to carry the national companies' offerings. On the basis of a report produced for the government by the Information Technology Advisory Panel, the decision was made in February 1982 to wire Britain 'as a matter of urgency' (ITAP, 1982). A government inquiry set up under the chairmanship of Lord Hunt reported in October 1982. It advised against severe restrictions on cable (as advocated by the existing broadcasting companies) in order to offer the public a wide variety of entertainment channels, but it declared against experiments such as pay-TV and in favour of the use of coaxial cables, with limited capacity rather than for optical fibre cables for carrying programmes (*Hunt Report* 1982). This would militate against any extended use of cable as a potentially interactive medium in the following decades. By contrast, the French government has decided to implement its own plans for the development of cable using fibre optics technology.*

The German Federal Republic is experimenting with a multi-channel cable in seven towns, including Berlin, in order to assess the economic and other effects for a possible Republic-wide implementation. Considerable controversy has arisen over the role of advertising and the effects on the press and existing broadcasting channels.

The transmission of cable services is often said to be 'narrow-casting' (as opposed to 'broadcasting'), because local cable companies will be able to carry local news, and be able to produce magazine-type services aimed at more narrow target groups than the broadcasting authorities. This will inevitably mean that new technology is likely to bring cable television into competition with newspapers and magazines and with the range of Teletext facilities now being provided.

In direct contrast to the supervisory system of Britain which may yet reduce the potentialities of cable, stands the Italian experience. Article 21 of the Italian Constitution calls for 'liberty of expression', and in accordance with this ideal the Constitutional Court made a decision in 1976 which effectively removed the monopoly enjoyed by RAI-TV, the state broadcasting body. There was a rapid expansion of private television operators using cable and conven-

*The British Government, in the Spring of 1983, in a White Paper, suggested that a 20-year Franchise would be issued to companies willing to adopt fibre optics technology, rather than one of 12 years.

tional broadcasting means – a mushroom growth of initiatives responding to the clear demand for alternative information and often growing in accordance with the needs of local communities. In some areas air waves became jammed and over-lapping of programmes occured inevitably, and even pornographic late-night programmes were offered! After a time, however, viewing patterns stabilised, giving Italians a wider choice of television. Advertising has been a source of finance for many of these operators.

4.3 Satellite Communications

In respect of the possibilities arising from the existence of satellites capable of transmitting national signals and signals from private operators across national boundaries, existing plans are for the transmission of the programmes of national radio and television broadcasts only. The technology could allow for the opening up of many forms of commerical and other interpersonal communications, yet at present such opportunities are likely to remain unexploited. The usual arguments in favour of control by government-controlled or regulated bodies is related to the concept of the scarcity of the airwaves. In the economics of television the problem of finance tends to strengthen political control over the organisations.

Services are financed by revenues from advertisements or from sponsorship. The commercial companies are financed by advertisements, but the number of advertisements shown is regulated, and a sharp distinction is drawn between programmes and advertisements. In Britain, the BBC is financed directly from the sale to the public of television licences, but although the government does not dictate how the revenue is spent, the amount of the licence fee is subject to parliamentary control. Broadcasting is essentially a public good because it requires real resources to produce and transmit programmes, but it is difficult to exclude consumers from receiving the service. This may lead to a free-rider problem for the operator, who requires revenue to run the service.

The launching of Britain's commercial satellite, likely to be Europe's first, has been estimated to cost over £150 million in a project led by a consortium of British Aerospace, British Telecom and GEC Marconi. In the space communications industry, government and industry are interlinked, with the Department of Industry, in 1982, supplying £54 million through the European

Space Agency to British Aerospace in order to produce Europe's largest communications satellite (L-Sat) due to be launched in 1986. In a space market likely to be worth £4 billion world-wide by 1986, the risks and pay-offs are high. This is true especially for the commercial communications satellites such as Western Union Corporation's Weststar V, which had made a profit of £20 million at the time of its launch in a project costing £80 million.

Government control (with respect to the licensing of airwaves, incidence of advertisements in television, and cost and collection of licence fees) allow subtle political control by government to be exercised over broadcasting. As Hood (1979) puts it: 'the official ideologies of both BBC and ITA (in Britain), lay claim to a degree of independence not born by the realities of the situation.'

Without government controls, the only impediments to large-scale broadcasting and reception across national boundaries are economic ones. However, the costs of transmission for any 'pirate', dubious operator or unapproved political group are likely to remain large in the conceivable future. A related problem is, of course, that of the financing of broadcaster-'approved' or 'unapproved' operations. The British government decided in the Spring of 1982 to allow the beaming of two satellite channels into the country by 1986. Partly to remove the problems of loss of control, and in order to encourage the development of the future cable TV industry, the BBC were allocated both channels. The signals from the satellite are to be collected by 'dish', and piped into homes by cable.

Already, voices are raised about the need to implement new safeguards and controls. The arguments for these controls often stress the need for preserving national culture. There is a powerful economic incentive to free the flow of information which conflicts with national governments' wishes to constrain it. A decision of the European Court in 1973 implied that radio and television transmissions should be treated as services, and therefore should come under the terms of Article 59 of the Treaty of Rome which calls for a free trade in services throughout the Community.

4.4   Press

There is much debate in industrialised countries concerning the effects of computerisation in the newspaper and publishing industries. 'Technological unemployment', and the deskilling of printing work, have been central to the controversy in the newspaper and publishing sectors. Another area of concern is with the effects of the new technology on the relative efficiency of the printing and newspaper industries. The new technology has not been equally, nor speedily, diffused throughout these industries, and therefore not all firms have benefited from increased productivity due to exploitation of the new technology. Financial failure has occurred because many firms have been late in adopting the new technology. In the case of newspapers, the problems created are not simply related to the displacement of jobs and to productivity, but also to important political issues.

The nature of the economics of newspaper marketing involves the receipt of revenue from cover price and advertising, and the simple equation of profitability equals market success hides a complex interaction between the two sources of funds, and their relationship to production costs in newspapers. Such financial problems facing newspapers fell within the remit of the Royal Commission on the Press which reported in 1977 (the McGregor Report), but which did not consider the introduction of new technology. The McGregor Commission came down firmly against any form of subsidisation by the state, maintaining a belief in the political neutrality of market forces.

The existence of a free market in newspaper publishing and hence in ideas presupposes at least three conditions which are rarely fulfilled in practice. There must be competition among a number of independent producers offering distinctive products; consumers' choice through purchase must determine these products; and entry must be open to all capable of making a profit without restrictions. But if financial failure results in an increased concentration of ownership, and a reduction in the number of newspapers available, then there may not be the dissemination of a range of political views. The problem is especially serious in Britain, and has engendered a debate concerning free market forces as opposed to state intervention.

Secondly, on the face of it, any arguments in favour of state subsidies to the press, or state intervention to support particular failing newspapers, lead to fears of state control. The free market may lead to a range of newspapers more closely aligned to the views of citizens as consumers (this is the argument, for example, for the deregulation of television broadcasts), but in no way guarantees the dissemination of a range of political views.

Thirdly, the condition that entry must be open to all entrepreneurs capable of making a profit without restrictions is rarely fulfilled, for the following two reasons: the high costs of establishment; and secondly, any newspaper requires sufficient revenues from readership and advertising to support its costs.

5   Politics of Information Technology

5.1   *Information Access*

It is unlikely that access to the information stored in data files will become freely available even though changes in technology are likely to lead to a continuing fall in the costs of information processing. 'Information', it has been said, 'is power', and unequally distributed information can lead to inequalities in the distribution of power, income and wealth. New information systems, including two-way cable services, if implemented, through a free market could avoid many of the problems of government control and create an 'information revolution'. On the other hand, they may create a mass of competing and expensive data banks for those who can afford them. Certainly, computer data banks on matters of commercial importance and data on private citizens already exist, and are capable of being linked together through instantaneous national or international networks by satellite.

The amount of information that can be so transmitted is large – of the order of 1 million 'bits' per second. Such storage and transmission capabilities raise fears about the right to privacy of individuals, as private organisations may be able to build up a profile of an individual for credit ratings or employment purposes based on a mass of possibly 'sensitive' sources. Doctors, for example, have been worried about the misuse of medical information stored on computer files. The right of the individual to control information about himself, or to have access to files kept about him, is fundamental to the operation of a society relatively secure from abuses of power by the state or large corporate bodies. The technology can allow fundamental changes in the balances needed to protect the individual right to privacy. MI5 in Britain have recently bought a computer system capable of holding information on every individual in the country. The British government admits to having 'speculative' files on 1.5 million people, many of whom have committed no criminal offence.

The dangers of closer state surveillance inherent in computer files have not gone unnoticed. In 1974, Sweden became the first country to legislate for data protection to protect personal liberty. By 1982, the US, Canada, and New Zealand had passed similar laws along with several European countries. The Council of Europe has formulated principles on the protection of privacy, and several other European countries are formulating proposals to enact legislation. But there remains the danger of data 'havens' becoming established in countries without effective legislation, or with no regulations at all. Organisations would then be able to store all personal files in these countries and access them at will via satellite.

A report published in Britain (Lindop, 1979) proposed the establishment of a statutory body with powers to inspect all state computer systems and to conduct spot-checks to ensure that personal data were handled with due regard for security and accuracy. The British government's proposals (1982) for legislation have been much weaker in scope. A Registrar will be appointed with whom all data banks will be registered which contain files on identifiable individuals. Individual access to check accuracy is to be allowed to all files except those relating to national security, police matters, medical data and other 'sensitive matters', such as social work reports. Yet in the case of private data banks, no legal codes of practice will be enacted, so that the onus is on the individual to investigate any suspected abuses, and then only civil action can be taken. Britain's proposed laws are far weaker than legislation already enacted in many other industrial nations.

Technical advances in telecommunications may make the collection of information for state computer files an easier matter. Phone-tapping used in industrial espionage, and politically as in the Watergate affair, has a long history as a means of state eavesdropping on political opponents. Mussolini, for example, used the technique regularly. However, until recently, phone-tapping, or monitoring and taping conversations, was always an 'obvious' form of interference. Engineers had physically to go to an exchange and install the taps, and although it is estimated that the number of phone-tapping exercises in Britain increased at a faster rate than the number of 'warrants' for taps for criminal investigations, the technical problem of installation remained a restricting factor on any moves towards mass surveillance. In Britain, 'System X' removes these technical limitations as it does not require engineers to install any monitoring devices; the operation is contained within microscopic switching devices and is controlled by computer. Lines can be monitored simply by changes in programming. As Campbell (1981) argues, 'this, if not regulated, can be turned into a tool of mass-surveillance far more readily than our present cumbersome electro-mechanical exchanges.'

5.2   Telecommunications

Technological advances have truly linked the world into a communications network, but Marshall McLuhan's (1964) statement that the world was becoming a 'global village' is far from accurate. For political and national cultural reasons governments maintain regulation of national communications networks. In considering the ownership and control of the media it is important to establish whether they reflect a true diversity of opinion, or whether they merely propagate the views of government, or the views of influential pressure groups.

Ten countries own 80 per cent of the world's telecommunications media while the entire Third World controls no more than 8 per cent. This has led to the criticism that western media represent an alien, richer culture, that they report inaccurately, and usually only failures of policy, and that they dominate with a type of 'information imperialism'. This has led to demands in the Third World for national control over the media to be an overriding principle in a new world information order as a necessary part of the struggle for economic and political independence. The problems of the inequality of ownership has led to these drastic demands which imply for many African, Asian, and Latin American countries, one reality: state ownership and control. Encouraged by the USSR, a meeting of Latin American governments called by UNESCO endorsed the principle of state involvement in national communications systems in 1976. The Third World and the western nations soon came into conflict over the call for a 'world information order' in a running dispute which nearly broke up UNESCO, and was only resolved after a special inquiry by the agreement to item 22 at a conference in Paris in 1978. (Item 22 is a draft of principles for the use of mass media. See Windlesham, 1981.) The result is a compromise which is not binding on any state, and many Third World nations have gone further towards implementing their own version of a 'world information order' in recent years. This is likely to cause future problems in world communications which will be exacerbated by the use of satellites which weaken national control.

We turn now to the politics of developments in the mass media. It has already been discussed how modern developments such as DBS, cable TV, and two-way interactive services by cable, may well revolutionise shopping and working patterns, and even lead to the disappearance of the printed media. Certainly, a wider choice of entertainment, and commercial and information services will become available by the 1990s, but the more radical effects on reading and social living patterns remain speculative. Most existing plans are for the transmission of national radio and television broadcasts only. Three reasons are given for this policy: first, the concept of the scarcity of airwaves; secondly, the fears expressed by countries of cultural dilution and penetration; and thirdly, governments' wishes to control centrally the mass media.

It has been argued that the mass media are not a means of communication so much as a means of distribution because they lack a two-way communications facility – advertising, news broadcasts, public information, and most entertainment lack feedback. Radio technology allows the possibility for every receiver to be modified into a transmitter, but although the right to receive is open, the right to transmit is closely regulated. The reason for this control, and a reason for central control of television companies, is that the spectrum of radio frequencies is a scarce national resource, so that the argument for regulation and close supervisory control by the state of television is based on the scarcity of the air waves which have to be rationed out. Modern cable systems overcome the scarcity problem since a large number of channels can be carried without interfering with each other.

'Overspill' problems in television broadcasting across national boundaries already occur, and are bound to increase in the future with the advent of direct broadcasting by satellite (DBS). European countries have already expressed concern about foreign broadcasts, and in 1977 an agreement between European states was signed to restrict satellite broadcasting, holding each country to the agreement to keep to a maximum of five channels; to limit broadcasts to national areas; not to transmit in foreign languages; and to ask for permission of another state if broadcasts aimed at its citizens are to be beamed within its boundaries. However, ultimately the problem is not so much a fear of cultural dilution, as one of a weakening of control at the political level. Certainly the possibility of the reception of DBS broadcasts, which have been transmitted from another country by satellite, threatens to break any country's national monopoly over its broadcasting.

Reception of broadcasts by satellite is more difficult to regulate than cable television reception, because all that is required by a household is a satellite-receiving 'dish'. As we have already noted, the British government, partly to remove the problems of loss of control and in order to encourage the development of the future TV cable industry, decided in 1982 to allow the beaming of two satellite channels by 1986. The signals from the satellites may be piped into homes by cable. The government chose to allocate both channels to the British Broadcasting Corporation (BBC), a measure designed to enable the central authorities to maintain their monopoly over consumers' broadcasting reception. However, difficulties of policing satellite reception exist. As the Secretary of the European Advertising Tripartite (Tempest, 1982) comments:

Conventional broadcasting by satellite presents vested interests in governments and national broadcasting institutes with the greatest threat, for it is eventually impossible to prevent the determined viewer from tuning in to foreign broadcasts which do not meet with his Government's approval just as it was impossible in the 1960s to prevent the popularity of pirate radios.

Although the initial cost of the purchase and installation of a receiving 'dish' is high, providing there is a large market scale economies and competition could lead to a reduction in the costs of DBS reception in the future. US experience has shown that DBS and cable TV are direct competitors. In the less regulated market DBS discourages cable subscriptions. In Britain, a major financial problem experienced by the BBC is that because it may not accept advertising as a source of revenue it has to find some other method of paying for expensive innovations which does not require either an unacceptably high licence fee or huge state subsidies.

5.3   The Press

The introduction of new technology into the printing and newspaper industries has proved to be extremely controversial. Public concern has been expressed about the effects of computerisation and new print technology on jobs, both in regard to employee displacement by high technology in newspaper production, and in the skills required for editorial work. The problems also raise important political issues.

The presentation of a balance of views and a means for maintaining freedom of discussion and communication of ideas in a pluralist society are often said to depend upon the existence of a 'free press' relatively untrammeled by government control (except through the laws pertaining to libel and obscenity). This principle is thought to be so important that it is incorporated, for example, in the first amendment of the US constitution. If financial failure of newspapers results in an increased concentration of ownership, and a reduction in the number of newspapers available, then there may not be the dissemination of a full range of political views.

Arguments in favour of state subsidies to newspapers lead to fears of state control and manipulation of the press. On the other hand, deregulation and a free market may lead to the elimination of newspapers with minority political views because of low advertising revenues. Distributors in the UK can be sued by an individual where he considers defamatory statements about him have been made in a newspaper, which discourages some distributors from handling minority interest newspapers. However, experience in France shows there is a way of maintaining reasonable diversity by legal intervention in the marketing and distribution of newspapers. Since 1947 it has been mandatory for distributors and newsagents in France to take any newspaper and to display it. Distributors have a sale or return arrangement with publishers, so that although they are under an obligation to carry all newspapers, the success of a newspaper depends upon its appeal to customers. In consequence France has a relatively large number of newspapers covering the whole political spectrum.

The future of the press does not look good in present political and economic circumstances. It is, however, not so much the introduction of new technology that may reduce the range of political views necessary to sustain the level of free speech essential to any liberal democracy, but, as is the case in the other forms of the mass media we have considered, it is the nature of the institutional, political and economic framework in which that technology is deployed that determines its social consequences.

6 Conclusions

Government and industry are closely involved in all of the modern developments in the field of communications we have discussed. Political discussion revolves around the alternative models of state control or regulation, and the possibility of reliance solely on the market as an allocative mechanism. Economists, supporting the 'invisible hand' idea of Adam Smith, have argued that prices play a role in market exchanges in such a way to limit the amount of information which needs to be transmitted between persons in order to coordinate activities. Control by the state or by large firms, it is suggested, implies a greater need to transmit information. But markets can and do fail to ensure that what is self-interest for the individual is 'rational' for society, and state control of a communications system, through regulation of commerce, and surveillence for internal security, may have the effect of reducing individual freedom. It may be that free-market economists expect too much of the price system, especially in relation to the problems of high technological societies discussed in this book.

Communications technology in no way prevents distorted or unreliable information from being transmitted. The dangers of misleading advertising and of persuasive claims with little truth are real, and require some degree of regulation, so that the majority of honest traders can signal in confidence, and the 'fly-by-night' trader is discouraged. But the economic and social consequences of distortions are not, many believe, nearly as dangerous as the use of communications networks for political propaganda by authoritarian or totalitarian states, such as occurred in Stalinist Russia, or in the propaganda machine of Goebbels in Nazi Germany, or in the fictional nightmare of George Orwell's *1984*. The dangers exist, even in liberal market economies, that 'technological imperatives' may lead to increasing control over economic and social life, with a resulting loss of freedom to the individual.

The continued control of the media by government, or by private organisations, limits their role and impact on society, especially their liberating potentialities. The application of the scientific and analytical approach to the area of communications, developing out of face-to-face communication, has generated an undreamt of possibility for instantaneous global contact between people. Yet face-to-face communication has been largely lost in modern communications technology. Early communications forms represented social interaction through messages: they contained feedback. Indeed, this two-way facility may be such an important feature of communications that it should be an essential prerequisite of any definition of it. The potentialities of true communications through the earlier technological marvel of radio were described by Brecht (1932) thus:

Radio must be changed from a means of distribution to a means of communication. Radio would be the most wonderful means of communication imaginable in public life, a huge linked system – that is to say, it would be such if it were capable not only of transmitting but of receiving, of allowing the listener not only to hear but to speak, and did not isolate him but brought him into contact. Unrealisable in this social system, realisable in another, these proposals . . . are, after all, only the natural consequences of technical development.

References

Brecht, B. (1932), 'Theory of radio', *Gesammelte Werke,* Bande 8 (Frankfurt: Hrsg. Vom Suhrkamp).

Campbell, D. (1981), 'Big Brother is listening', *New Statesman,* Report no.2.

Carter, C.F. (1977), *Report of the Post Office Review Committee,* Cmnd 6850, (London: HMSO).

Data Protection (1982): The Government's Proposals for Legislation, Cmnd 8539 (London: HMSO).

Economist, (1982), 'The wiring of Britain', 6 March, p. 11.

Gerbner, G. (1967), 'Mass media and human communications theory', in McQuail, D. (ed.), *Sociology of Mass Communications* (Harmondsworth: Penguin), pp. 35–58.

Gould, J., and Kalb, W.L. (1964) (eds.), *A Dictionary of the Social Sciences* (New York: Free Press).

Hood, S. (1979), 'The politics of television', in McQuail (ed.), *op. cit.,* pp. 406-34.

Hunt, Lord (1982), *Report on Inquiry into Cable Expansion and Broadcasting Policy,* Cmnd 8679 (London: HMSO).

Information Technology Advisory Panel (1982), *Report on Cable Systems,* Cabinet Office (London: HMSO).

Lindop, N. (1979), *Report of the Committee on Data Protection,* Cmnd 7341 (London: HMSO).

McGregor, O.R. (1977), *Report of the Royal Commission on the Press,* Cmnd 6810 (London: HMSO).

McLuhan, M. (1964), *Understanding Media* (New York: McGraw-Hill).

Tempest, A. (1982), 'The witches' brew, advertising, satellites, and the European institutions', *Journal of Advertising,* vol. 1, pp.143-55.

Williamson, O.E. (1976), *Markets and Hierarchies: Analysis and Antitrust Implications* (New York: Free Press).

Windlesham, Lord (1981), *Broadcasting in a Free Society* (Oxford: Basil Blackwell).

11 Scientific Management and Work

PETER WHEALE

1 Introduction

In the late nineteenth century the systematic application of scientific knowledge to industrial processes gathered momentum. In the 1880s rapidly expanding electrical and chemical industries emerged in the major industrial countries. This successful growth of science-based industry reinforced the scientific ethos prevalent in these countries: spiritual truth apart, 'scientific' knowledge was becoming the only 'legitimate' knowledge acceptable to industrialised society.

Many scientifically-minded people, particularly electrical and chemical engineers, became employees of the new and expanding industrial enterprises. As this new breed of employees worked their way up to managerial and executive positions within the science-based corporations, they came to identify the advance of modern technology with the advance of these corporations. The application of science to the arts and crafts had become the professional habit of the engineer and the operating policy of the industrial firm. Noble (1979) observes:

Even in his strictly technical work the engineer brought to his task the spirit of the capitalist. His design of machinery, for example, was guided as much by the capitalist need to minimize both the cost and the autonomy of skilled labour as by the desire to harness most efficiently the potentials of matter and energy.

It was in this 'scientific climate' that the engineer Frederick Taylor, working in the US, began in the 1880s to combine the basic behavioural assumptions of classical economic theory with the formal discipline of engineering to create a school of thought which eventually became known as scientific management. Taylor sought to introduce a systematic division of labour, work performance governed by rules scientifically derived, and a system of written instructions and communications.

In this chapter we examine the procedures of scientific management. Section 3 considers the alienation of the worker in industrial societies from work: the view is examined that the scientific management approach, incorporated today into much of the new technology based on developments in electronic and computing science, perpetuates a form of social relations between corporate management and the workforce which tends to alienate the latter from the former. Section 4 discusses occupational hazards to health, and the possibility that the high incidence of accidents and industrial illnesses is due to the 'scientisation' of work. And we conclude with a brief account of the problem of 'technological unemployment'.

2   Scientific Management

The rise of science-based industries reinforced the changes in business organisation which had begun to take place in the nineteenth century. The division of labour was already highly developed when Adam Smith published his account of pin-making in the *Wealth of Nations* in 1776, but with increased mechanisation work became even more specialised. Mechanisation provided mass-produced products and industry greatly expanded. The formation of joint stock companies enabled large amounts of capital to be invested. This expansion of the business enterprise required of managers (sometimes engineers themselves) greater competence in organising labour and capital than most of them had.

Sole control by entrepreneurs of their enterprises was challenged by increased trade unionism. In the United States, for example, trade union membership grew by about 500 per cent between 1897 and 1904. (See Bendix, 1956.) In their attacks on the trade unions, employers came to make their own absolute authority within the plant a central tenet; the compliance of the workers became ideologically of far more importance to the employer than the value of workers' independence and initiative. US employers of the time vigorously asserted their authority and strength in the open shop campaign.

Scientific management, sometimes called Taylorism or task-management, was believed to hold great promise since it claimed to promote material wealth and social harmony at the same time; disputes were to be settled rationally by appeal to 'objective' criteria. It claimed to raise productivity and lower production costs by the 'scientific' study of the organisation of the factory, office or shop and the operations of the individual workers. Although Taylor was the first to systematise a theory of work efficiency, others had in some ways anticipated him, notably, Charles Babbage (1792-1871) who belived in the labour-saving quality of machines, and in *The Economy of Machines and Manufacturers* (1835) had argued that the division of labour can be applied to mental as well as mechanical operations. Just as Adam Smith had seen the traditions and customs of the eighteenth century as undesirable barriers to efficient markets, so Taylor perceived traditional manufacturing procedures as impeding rational production systems and thereby reducing productivity. He therefore attempted systematic analysis of the labour process and the division of labour, developing a number of principles. Littler (1978, 1982) has summarised Taylor's principles as follows:

1. A general principle of maximum fragmentation which decomposes work into its simplest constituent elements or tasks.
2. A divorce of planning and doing work, thus removing as far as possible any discretion within the work to be performed.
3. The divorce of direct and indirect labour which embodies the principle of task control. Here a planning department was envisaged to plan and coordinate the manufacturing process.
4. Minimisation of skill requirements (i.e. the conception and execution of techniques) and job-learning time.
5. Reduction of material-handling to a minimum. A mechanism was sought for providing an efficient monitoring system over factory work with the use of time and methods study which was used to find the one best way of doing a job.

Taylor's system involves a scheduling procedure whereby workers complete job-cards or time-sheets detailing the work to be done and enabling management planning departments to monitor this information in order to determine effort levels and compare performance. The mechanisms of control advocated by Taylor and his associates involved not only control over task performance, but the means towards perpetuation of that control. In practice, the

Taylorian model involved staff-line organisation, planning departments being added to the previously existing command structures of organisations. Taylorism embodies the transfer of all the traditional knowledge, job discretion and skill of the workforce to management in the forms of these planning departments. Organisational authority can be described in terms of 'span of control', and is designed on a pyramidal authority structure. (An individual's 'span of control' is the number of people directly responsible to him.)

The image, or 'model of man', implied by scientific management is that of the worker who is by nature lazy, indifferent to the goals of the organisation, and hostile to authority. A system of authoritarian controls is therefore considered necessary for obtaining satisfactory work performance. Mass production often led to an individualisation of work roles and to the replacement of group by individual incentives which facilitated greater management control over the individual worker (Kelly, 1982).

A somewhat broader conception of the worker was held by some of Taylor's early associates, notably Gantt and Gilbreth, who were more aware of the social and psychological aspects involved in motivating work performance. The modern management movement, which began at the turn of the century, incorporated these social and psychological aspects of the worker, and advocated less punitive and authoritarian procedures than strict Taylorism. However, the manipulation of the worker to increase his productivity (human engineering) though more subtle than in Taylorism, was the primary objective of modern management.

Elton Mayo and his colleagues at the Harvard Business School investigated informal group norms of behaviour in the Hawthorne Studies (1924–31) – work-performance studies carried out at the Hawthorne works of the Western Electric Company near Chicago. Their findings inspired what was later known as the human relations school of management thought. Taylor was, however, well aware of the existence of work groups, and the way workers established informal and less productive norms of output performance (i.e. lower than was technically feasible), which he referred to as 'systematic soldiering'. Indeed, his whole task-management system was designed to destroy this informal power assumed by workers. By standardising each set of job operations, Taylorism was deliberately designed to break the power of work-groups, and by linking pay directly to productivity through piecework schemes, to appeal to individual ambition, so tending to 'atomise' the workforce. Furthermore, as Braverman (1974) has pointed out, the subdivision of work into separate tasks, with different rates of pay attached to each set of tasks, enables employers to reduce their overall payment to labour, thus increasing profits.

Theorists of modern management and the human relations school of management perpetuate Taylor's human engineering approach to worker motivation. Job design continues to reflect the profit objective, and notions such as 'job enrichment', through job rotation, or 'quality control groups', are used by management to elicit increased productivity, and not primarily to make tedious work less tedious. From the outset, modern management industrial relations programmes aimed at higher efficiency: company doctors reduced absenteeism, and better heating, lighting, ventilation and canteen food were believed to improve productivity.

Taylor made such a great impression on the industrialists of his day that a society was formed to disseminate his ideas, and his *Principles of Scientific Management* (1911) was translated into a dozen languages. Lenin, in the new communist Russia, expressed enthusiasm for the rational production methods and scientific job analysis incorporated in scientific management; and the Japanese were also eager to incorporate the procedures into their new industries; their branch of the Taylor Society was one of the first to be formed. Dissemination of Taylor's ideas also occurred in Germany and Italy, and communist China later assimilated task-management into its Cultural Revolution.

Management strategies have varied greatly in practice, and as a manufacturing system Taylorism has only been adopted in a piecemeal fashion. Indeed, foremen and sometimes managers often successfully resisted the introduction of time-and-motion studies and the related organisational changes recommended. In Britain, for example, in engineering, where unionisation was well developed, resistance to Taylorism was mostly successful.

Braverman's study, *Labor and Monopoly Capital* (1974), has been rightly criticised for over-emphasising the effects of Taylorism in the process of deskilling work (Wood, 1982). In discussing the degradation of work in the twentieth century, Braverman romanticises the pre-industrial craftsman. Taylor's system of scientific management merely fitted into an ideology of scientisation to which many social, economic and technological factors were predisposed.

Nevertheless, the wide dissemination of scientific management ideas throughout the industrialised world helped ensure that the scientisation of work became a part of management consciousness.

3 The Alienated Worker

The concept of alienation was developed by Marx as part of his critique of capitalist society (Marx, 1844). According to Marx people are alienated in capitalist society from nature, from themselves, from each other, and from their community. Alienation from work arises because workers are not in control of the 'means of production' (technology), and secondly because the employer 'buys' their labour power just as he would buy any commodity on the market. Marx believed the first divorces workers from the products of their labour, and the second dehumanises workers.

Control of workers, whether by owners or managers, has stimulated many studies of people at work, using the concept of alienation as a central theme. Dubin (1956) found that in a sample of industrial workers in the United States, three out of four workers did *not* regard their work as a 'central life interest'. Continuous assembly-line production and other forms of automation have been considered as alienating work – work is reduced to an unsatisfying chore, thus destroying the worker's self-image. Several work studies have found that work which was machine-paced, that is, dictated by the speed of the machinery used, was especially disliked. Goldthorpe *et al.* (1968), in a study of a British factory, found that car workers had an *instrumental* view of their work – they were prepared to accept deprivations at work in exchange for high pecuniary rewards. Emphasis in this and other similar studies is given not only to the influence of technology in shaping workers' attitudes at work, but also to the attitudes and expectations which they bring with them to the factory. The concept of skilled labour can be considered as not only the combined conception and execution of techniques, but also the incorporation of worker control over the labour process and the status of the job (Beechey, 1982).

Woodward (1958, 1966), in a study of manufacturing firms, concluded that technical methods were the most important factor in determining the organisational structure. The attitudes and behaviour of management and supervisory staff, and the nature of industrial relations in the firms, appeared to be closely related to the technology they used. Firms were divided into three classes according to their techniques of production: small-batch and unit production, large-batch and mass production, and process production. In firms at the extremes of the scale, relationships were, on the whole, found to be better than in the middle ranges. Woodward's work indicates that assembly-line production was the most alienating.

Figure 11.1 depicts the likely incidence of alienation of the

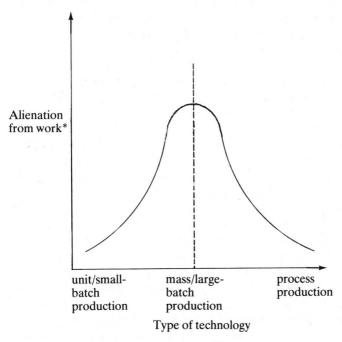

Figure 11.1: Alienation from work

Source: Woodward, J. (1958), *Management and Technology* (London: HMSO).

* Alienation from work has been measured by recording the attitudes of workers towards their jobs, 'strike proneness' may also be considered as a proxy measure alienation of workers from their work. Prais (1978) found that there was a direct relationship between plant size in manufacturing and 'strike proneness'.

workers from the technology using Woodward's classification. The point to note here is that small-batch or unit production is more likely to involve discretion in the job on the part of the worker than mass production. On the other hand, process production techniques often require the ministrations of technical experts. In both these types of technologies the workers are likely to be more involved and thus less alienated from their work than those working under mass production (assembly-line) technology.

Woodward's results indicate that different technologies impose different kinds of demands on individuals and organisations. Certain other studies (see, for example, Hickson *et al*, 1969) favoured organisational size as the most important determinant of organisational structure, but as mass production and extensive use of automated machinery require large organisations, there is no substantial contradiction of Woodward's work. Organisational structures are influenced by factors other than simply technical ones. Arguments based on technological determinism are likely to be found wanting when actual organisations in different industries are closely examined. 'Span of control', Woodward found, reaches a peak in mass production, and the numbers of levels of authority in the management hierarchy tend to increase with technical complexity. Although production control becomes increasingly important as technology advances, the administration of production ('the brainwork of production') is most widely separated from the actual supervision of production operations in large-batch and mass-production firms, where the newer techniques of production planning and control, methods engineering and work study, are most developed.

Blauner (1964) drew similar conclusions from his study of the printing, textile, chemical and car industries in the US. Alienation from work, he argued, is at its highest, and the workers' freedom at its lowest, in the assembly-line industries. Although he considered the workers' control over the work process had been increased by the introduction of continuous-process production, he expressed reservations concerning the continuance of this trend in the future as automated technology is introduced. The challenge of numerically-controlled (NC) machines in the 1980s is to retain the versatility of the machine so as to give full rein to the skill (the combined conception and execution of techniques) of the operator.

The application of science to industrial methods of production tends to treat labour as an object: labour and capital become mere inputs to the production function, and the labourer becomes little more than an appendage to his or her machine. The activity of the workers becomes a means to produce goods for profit, and never an end in itself. Blauner (1964) considered the worker is more likely to be used as an 'object' when he lacks control in the work situation, and when his role is so specialised that he becomes merely a 'cog' in an organisation. Taylor and his associates, in assuming the mantle of science, had propounded a universal theory of work: a 'one best way' to perform a job; a blueprint for efficient management; and a 'rational', and thus detached, attitude to labour. The division of mental and manual labour based on time-and-motion studies of each job in isolation, transfers all mental and discretionary aspects of a job to the managerial staff.

Control by computerisation can now be achieved by means of mini-computers and computer terminals linked to a central system. In this way a control centre can be connected to shop-floor terminals and to word-processors in offices. Systems analysis, operations research techniques, quality control systems, and statistical and economic analysis have been revolutionised by the new information technology. The use of robots and microprocessor-controlled machines has led to a restructuring of the work to be performed (Noble, 1978). Traditionally, industrial laboratories, design offices and administrative centres have been the enclaves of conceptual and planning work. Job satisfaction was usually attained by employees working in these areas because they were likely to be interested in their work, which demanded skill, expertise and judgement from them. Computerisation, it is argued, is deskilling work, and scientific management techniques are now being applied to jobs not formerly considered amenable to them. Cooley (1980) suggests that the computer is acting as a Trojan horse for Taylorism: the results, he argues, are eliminating human judgement, job discretion and the subjective and creative nature of work which was formerly stimulating and satisfying.

Algorithmic methods of problem solving require predictability, repeatability, and quantifiability, which imply the elimination of unique human judgement and subjective creativity. Cooley (1980) uses computer-aided design (CAD) as an illustration, arguing that the quantitative information designers amass before making qualitative judgement is extremely complex, and that the crude introduction of the computer into the design process by management, to divide the qualitative from the quantitative aspects of a job, results in a deterioration of design quality. Algorithmic methods reduce the decisions left to the operator of the system to routine choices between fixed alternatives. Similarly, computer systems used to systematise building design, by arranging predetermined elements on a visual display unit (VDU) in order to produce different building configurations, limit creativity in the job to choosing how the different elements will be disposed, rather than considering, in an open-ended way, the types and different forms and materials that might be used. Similar computer systems are now used to draw maps and charts. This tends to deskill the work of the cartographer and downgrade the quality of his work.

It is ironic that cybernetic techniques are now used in some firms to help stimulate individual and work-group creative thought, when it is the very philosophy behind cybernetic control which has aided transfer of the discretion in work from the worker to management. Indeed, creativity techniques such as brainstorming and synectics may be seen as 'treatments' for people in a society suffering from general repression of its imaginative and creative abilities (Wheale, 1978).

Information and data control systems are used to control and plan both the production and financial aspects of business. These systems not only automate assembly-line work, control robotic machines and stock movements, diagnose machine faults and the like, but are now automating office work as well. Word-processors, for example, have deskilled typing and clerical work: complex editing, revision of lay-out, automatic sorting, electronic transmission of mail and laser printing at very high speeds are now widely employed. All these innovations increase productivity at the expense of labour. For those staff who remain, the intensity of work often increases, contact with other staff is greatly reduced, and management is able to monitor staff activities much more closely. As Braverman (1974) has observed, the fact that these machines may be paced and controlled according to centralised decisions, and that these controls are in the hands of senior management, is of as much interest to management as the fact that they increase the productivity of labour.

Alternative job designs do exist and could often replace alienating work. The interactive use of computers, where the operator is able to alter the procedures and use judgement with the help of a computerised machine, retains the skill of the operator. In designing which is computer-interactive, the procedure can give full range to the discretion of the individual in assessing quantitative results and selecting a solution. As we pointed out in Chapter 3, although mathematical analysis enters into engineering in an indispensable way, good engineering requires experience and judgement not amenable to scientific analysis alone.

Job design is best where it is liberating and life-enhancing and this can never be achieved where all the personal autonomy of the workers is assumed by management. Computers in factories can encourage organisational decentralisation, and increased automation can be installed in such a way as to raise the level of skill and responsibility of workers. (See, for example, Blau *et al.*, 1976.)

4 Occupational Health Hazards

Occupational health risks are a symptom of the 'scientisation' of work, where labour is treated as an object rather than a human enterprise. Industrial poisonings, accidental infections, eyestrain, exposure to pollutants such as asbestos fibres, dust, lead and toxic fumes, are common in modern industry. Cox (1980, 1982) has studied human stress at work and found it to cause health problems: he argues that stress at shop-floor level may be greater than that at higher levels of the organisation, since lack of control over the job, and boredom through repetitive jobs or machine-paced operations may cause greater mental stress than work which in many ways is more demanding.

The urban family, Weinbaum and Bridges (1976) argue, is a unit of consumption, an 'internal market', such that household labour can be degraded in a similar way to that of wage labour. Indeed, housework in an isolated environment can be stressful, and accidents from cooking, and skin diseases contracted from cleaning fluids, are common.

When the British Committee on Safety and Health at Work (1972) reported, its opinion was that the toll of death, injury, and ill health was unacceptably high. The Committee, chaired by Lord Robens, argued that apathy was the greatest problem, and that reform should be directed towards enhancing effective self-regulation by employers and employees, rather than towards more detailed statutory legislation. Industrial health science has concentrated on the individual worker, and not on the technological environment in which he or she is placed. The tendency has been to attribute blame to individuals for the incidence of accidents and ill health at work: individuals are blamed for apathy towards safety, accident prone-ness, carelessness about protective clothing, or genetic suscepti-bility. Clutterbuck (1976) suggests that we should try to re-engineer technology rather than people. He argues that the wearing of protective clothing should be de-emphasised, as should the use of safety spectacles, ear-muffs, barrier creams, and such like, and instead we should focus on attempts at reducing noise levels, preventing the handling of dangerous chemicals and exposure to bacteria and so on, to provide a generally much less hazardous working environment than the one which currently prevails.

5 Technological Unemployment

Just as product and process innovations revolutionised the electrical and chemical industries in the late nineteenth century, so innova-tions in electronics, petrochemicals, nuclear engineering, transport and mineral extraction have wrought a fundamental restructuring of today's industrial economies. As we observed in Chapter 1, there has been a decline in the numbers employed in the primary sector of the economy (e.g. in agriculture, fishing, mining), and an increase in the secondary sector. In recent years there has been a fall in the number of workers engaged in the secondary sector of the economy (e.g. in manufacturing, utilities, construction), in favour of the tertiary (or services) sector. Furthermore, as the spread of infor-mation technology (the term used to describe the interrelated applications of microelectronics, telecommunications and com-puting) displaces typists, clerks, administrators and managers, so the growth of employment in the service sector of the economy is itself checked. Evidence of this is given by the recent computerisa-tion of handling activities such as insurance, banking, accounting, legal services and retail distribution. As this new technology is capital-intensive, it displaces labour, causing unemployment. Figure 11.2 shows recent increasing trends of unemployment in industrial countries.

In industrialised countries, because women tend to be concen-trated in the service sectors of industrial economies, and in un-

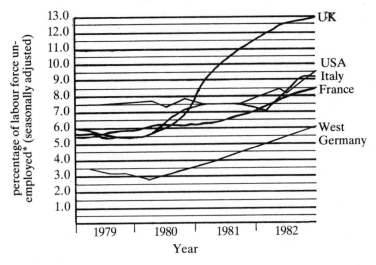

Figure 11.2: Trends in Unemployment
*National unemployment figures have been adjusted to the International Labour
Organisation's definition to preserve comparability between countries.
Source: OECD Statistics, 1982

skilled or semi-skilled jobs in manufacturing industries, the impact
on female unemployment due to the new technology is even greater
than that on men's jobs. Office automation brings about fundamental changes in the structure of employment in the workplace:
the ratio of clerical workers to professional staff falls, and the gap in
technical skills often increases. Office workers using the new technology tend to be less mobile in their work than before; the range of
skills required of clerical staff is often lower; verbal communication
between office workers, and their control and responsibility over
their work, are reduced. (See Huws, 1982).

These changes, particularly the decline of the secondary sectors
of industrial countries, have prompted some economists to refer to
a process of 'deindustrialisation' (see Blackaby, 1979). Deindustrialisation is not simply the result of capital-intensive technological
innovation, it is caused by a combination of factors, and is more or
less severe, depending on the presence and extent of these factors in
an economy.

The UK provides an example of a severe case of deindustrialisation. It is argued that the expansion of the public sector has taken

manpower from the private sector of the economy, and caused a 'crowding-out' of finance for the private sector. This shortage of finance has helped to push up the rates of interest for loanable funds. In the UK a relatively high exchange rate for sterling has prevailed in recent years due to the value of North Sea oil and gas reserves. This in turn has meant that the prices of exports from the UK have often been uncompetitive in overseas markets. The hope that the UK could rely on an expanding service economy which could market its services (e.g. insurance, banking) to provide increased export earnings ('invisibles') has not been fulfilled – in fact the UK's share of traded services has been falling in recent years. Many economists believe that the present predisposition of western governments to pursue anti-inflationary economic policies is likely (at least in the short-term) to constrain growth and further increase unemployment. The question governments ought perhaps to consider, is whether higher levels of unemployment are likely to remain a long-term feature of advanced industrial society. (See, for example, Hines and Searle, 1979.)

Two possible scenarios have been suggested for the main direction of change: (i) the 'soft' transition (a post-industrial society in which a greater proportion of non-material wants and needs is given a greater opportunity for self-articulation and self-organisation); and (ii) the 'hard' transition (a transition towards hyperindustrialisation – new, large-scale replacement technology and more control by state regulation and government agencies). Mensch (1979) argues that both capitalist and socialist countries have favoured 'hard' transition – i.e. preferring hyperindustrialisation to new expansionary technologies, those likely to create economic growth and higher levels of employment, and social innovation programmes.

Mensch believes industrial society is

largely alien to a 'soft' transition towards more of a 'participative economy' in which a larger share of the value-adding activities is nonhierarchically organized, mutually agreed upon, and voluntarily participated in as a leisure-time occupation of emancipated people.

Instead of our creating more social innovations – organisation reforms, new patterns of usage and of leisure – industrial stagnation is countered by yet more industrialisation. Encouraged by the patents system which rewards technical invention, the inventor is motivated to concentrate his creative energies into areas of technology where material technical progress can be made. The optimistic view of this process is that eventually the new technical innovations will create new industries which will supply wholly new consumer markets, and 'full' employment will be achieved again.

The bias is, however, toward rationalisation in industry and industrialisation of service delivery. These developments exert a much stronger pressure on the allocation of resources than the 'soft' transition to self-employment and self-organisation of exchangeable services and social contributions. In the commercial marketplace it usually pays to mass-produce products, thus displacing diversified production. The craftsman and the skilled worker are continually being dispensed with.

Noble (1979) argues that the use of modern technology is integrated with the rise of corporate capitalism. This implies that the social relations of science and technology have assumed a form whereby labour is alienated from those who own and control industrial enterprises. Furthermore, increasing competition and the search for greater profitability in the face of a declining world market for traditional products is intensifying industrial relations conflicts and creating, as Figure 11.2 shows, a huge pool of unemployed. Many thinkers consider that the choices open to advanced industrial society are political and ideological as well as technical. Cooley (1980) writes:

As we design technological systems, we are in fact designing sets of social relationships and as we question those social relationships and attempt to design systems differently, we are then beginning to challenge, in a partial way, power structures in society.

Pressure groups, such as trade unions and employers' organisations, should, we believe, discuss together the impact of rapid technological change on employment levels, skill content and the environment of work. These groups should assume the responsibility of co-operating together in establishing mechanisms and procedures to allow the process of technological change to take place in such a way, and to be of such a nature, as to increase the democratic freedom and general welfare of all people in society.

References

Babbage, C. (1835), *On the Economy of Machines and Manufacturers* (Fairfield, N.J.: Kelly, 1971).

Beechey, V. (1982), 'The sexual division of labour and the labour process: a critical assessment of Braverman', in Wood, S. (ed.), *The Degradation of Work* (London: Hutchinson).

Bendix, R. (1956), *Work and Authority in Industry* (London: John Wiley).

Blackaby, F. (ed.) (1979), *Deindustrialisation* (London: Heinemann).

Blau, P.M. *et al.* (1976), 'Technology and organisation in manufacturing', *Admin. Science Quarterly,* vol. 21, no. 1, pp. 20-40.

Blauner, R. (1964), *Alienation and Freedom* (Chicago: University of Chicago Press).

Braverman, H. (1974), *Labor and Monopoly Capital* (New York: Monthly Review Press).

Clutterbuck, C. (1976), 'Death in the plastics industry', *Radical Science Journal,* no. 4, pp. 61-80.

Committee on Safety and Health at Work (1972), *Safety and Health at Work: Report of the Committee* (the Robens Report), Cmnd 5034 (London: HMSO).

Cooley, M. (1980), 'Computerisation – Taylor's latest disguise', *Economic and Industrial Democracy,* vol. 1, pp. 523-39.

Cox, T. (1980), 'Repetitive work', in Copper, C.L. and Payne, R (eds.), *Current Concerns in Occupational Stress* (Chichester: Wiley).

Cox, T. *et al.* (1982), 'Repetitive work, well-being and arousal' in Murison, R (ed.), *Biological and Psychological Basis of Psychosomatic Disease* (Oxford: Pergamon).

Dubin, R.J. (1956), 'Industrial workers' worlds: a study of the central life interests of industrial workers', *Social Problems,* vol. 3, pp. 131-42.

Goldthorpe, J.H. *et al.* (1968), *The Affluent Worker: Industrial Attitudes and Behaviour* (Cambridge: Cambridge University Press).

Hickson, D. *et al.* (1969), 'Operations technology and organisation structure: an empirical reappraisal', *Administrative Science Quarterly,* no. 14, pp. 378-97.

Hines, C. and Searle, G. (1979), *Automatic Unemployment*

(London: ERRP).

Huws, U. (1982), *Your Job in the Eighties: A Woman's Guide to New Technology* (London: Pluto Press).

Kelly, J.E. (1982), *Scientific Management, Job Redesign and Work Performance* (Academic Press: New York).

Littler, C.R. (1978), 'Understanding Taylorism', *British Journal of Sociology,* vol. 29, no. 2, pp. 186-202.

Littler, C.R. (1982), *The Development of the Labour process in Capitalist Society* (London: Heineman)

Marx, K. (1844), *Economic and Philosophic Manuscripts of 1844* (Moscow: Progress Publishers, 1974), pp. 67-83.

Mensch, G. (1979), *Stalemate in Technology* (Cambridge, Mass.: Ballinger).

Noble, D.F. (1978), 'Social choice in machine design: the case of automatically controlled machine tools, and a challenge for labour,' *Politics and Society,* vol. 6, nos. 3 and 4, pp. 313-47.

Noble, D.F. (1979), *America by Design* (New York: Alfred A. Knopf).

Prais, S.J. (1978), 'The Strike proneness of large plants in Britain', *Journal of the Royal Statistical Society,* Series A, vol. 141, 3, pp. 368-84.

Taylor, F. (1911), *Principles of Scientific Management* (New York: Harper & Row, 1947).

Weinbaum, B. and Bridges, A. (1976), 'The other side of the pay check: monopoly capital and the structure of consumption', *Monthly Review,* vol. 28, no. 3, pp. 88-103.

Wheale, P.R., (1978), 'Creativity and cybernetics', *Creativity Network,* (MBS), vol. 4, no. 2, pp. 26-39.

Wood, S. (ed.) (1982), *The Degradation of Work?* (London: Hutchinson).

Woodward, J. (1958), *Management and Technology* (London: HMSO).

Woodward, J. (1966), *Industrial Organisation: Theory and Practice* (London: Oxford University Press).

Part III
QUESTIONS OF CONTROL

12 Science and Technology as Means of Control

CHARLES BOYLE

1 Introduction

The idea of technocratic control conjures up images of a world ruled by robots, and of latter-day Frankensteins tyrannised by the monsters they have brought into being. Yet these fictitious visions reflect real anxieties and fears. Science and technology have always been linked to the idea not just of knowing the natural world, but of dominating and exploiting it. Such a project of domination has not always met with universal or unqualified approval. The Greeks, for example, believed it betrayed overweening human arrogance, and they feared that this hubris would bring down the wrath of the gods.

In this chapter we explore the potential of science and technology for control, tracing it back to the scientific world-view that emerged in the seventeenth century. In Section 3 we outline some of the ways in which control is exercised, not only over the physical environment and processes of production, but also over man himself; and Section 4 examines views about the price we pay for such control.

2 The Scientific World-View and its Critics

In Chapter 2 we suggested that in seventeenth-century Europe a revolution occurred which brought about in the minds of a small but increasingly influential minority a new world-view, a new perspective on nature, the universe and man's place in it. This scientific revolution, and the world-view it entailed, were to lead eventually to enormously increased powers of control over nature. There have been considerable modifications to the seventeenth-century scientific world-view, but they are much less striking than the break with the medieval tradition which preceded it. Its full implications, we shall argue, are still being systematically worked out, both intellectually through the theoretical development of the various sciences, and practically through the various projects of technology.

We cannot fully appreciate the changes occasioned by the new perspective without thinking of earlier world-views, for example, those that form the background to the ideas and scenes depicted in the Old Testament and in Greek tragedies. It is important to realise how remote from us, not only temporally but also conceptually, are those worlds, with modes of thought, vocabularies, mental orientations and frameworks – in short, intellectual tools – very different from those of today.

Which few outstanding features of the complex pre-scientific way of looking at the world might be selected (accepting the inevitability of over-simplification) to point up a later contrast? Most importantly, both physically and spiritually, man was accorded a central place in the universe – a *human*-scale universe. The sun and the planets and the stars in their sphere moved round the earth; and on the earth all living creatures had their appointed places in the great chain of being (Lovejoy, 1965). Man's place set him apart as superior, the special creation of God, though still far inferior to the angels; and each human being had his or her proper position, as ordained by God. In addition there was a profusion of angels, ghosts, spirits and supernatural beings in a rigid hierarchy, many of whom had special magical powers. The sacred was distinguished sharply from the profane, and religion exercised a strong power on the minds of men. The authority of the Bible, the Church Fathers, and a small number of philosophers such as Aristotle, was supreme. Little attention was paid to the study of the natural world; there was an emphasis on quality rather than quantity, and explanations for phenomena were sought in terms of potentialities, purposes and final causes. By 1600 Copernicus had challenged the geocentric model of the universe, the Reformation had broken the unity of the Christian Church, and Aristotelianism had given way in some circles to variants on the ideas of Plato. But many of the earlier values and attitudes persisted, as indeed they did also throughout the seventeenth and even the eighteenth centuries among the very large sections of the population who were untouched and unmoved by the growth of science.

It was against this background that the new scientific perspective was established. One may characterise it in many ways, but for convenience of discussion here we separate out four aspects which we label, somewhat loosely, the *quantitative*, the *analytical*, the *experimental* and the *totalitarian*. These aspects are all interrelated ways of looking at a whole, fairly rigid, body of belief.

The new world-view was *quantitative* in that it placed mathematics at the very centre of its concerns. Not only was mathematical knowledge believed to be the only knowledge that was unchallengeable, and therefore a proper basis or language for investigations, but the old idea stemming from Pythagoras was taken up, that behind the surface of things there is ultimately a mathematical reality. Thus the belief grew that proper insight and understanding could be gained only by looking for regularities in nature that reflected the regularities of mathematics and could be described in mathematical terms. Classical mathematics gives great weight to regular functional relationships between variables; there is a tendency to turn a blind eye to points that lie far off the smooth expected graph, and thus there is an emphasis on regularity at the expense of the unique, one-off phenomenon.

Naturally, this emphasis on mathematics led to a very selective approach. A quality which could be defined or redefined in such a way as to render it capable of being quantified was elevated to a position of central importance; that which was not susceptible to measurement was regarded as ontologically inferior. As an example, we can consider the separation of 'primary' and 'secondary' qualities of an object by Galileo and Descartes. The primary qualities such as shape, size, position, motion, could be represented by numbers and had more reality and importance than the secondary qualities such as colour, smell, taste, which were held to be situated in the perceiver of the object rather than in the object itself. Similarly, perceived space and perceived time were replaced by abstract space and abstract time which were measurable.

The new world-view was *analytical* in the sense that it introduced most successfully the technique of considering the world as composed of separable elements each of which could be extracted, idealised and isolated from all but one or two specified influences. In mechanics one studied point masses interacting with each other, but removed from the attractions of the rest of the universe. This is essentially an *atomistic* approach, and is at the opposite pole to a world-view in which every object is regarded as closely related organically to every other object, and in which every event influences every other.

An important further feature of the analytical approach was the use of the concept of mechanical causality to link events and to define what was acceptable as an explanation. Descartes, for example, was deeply impressed by the mechanical, mathematical regularity of clockwork, and developed clockwork models and metaphors to describe both man and the universe. Teleological explanations (in terms of purpose) were rejected in favour of descriptions listing chains of cause and effect. Questions beginning 'why?' were rephrased to begin 'how?'. This was another very important break with the medieval, Aristotelian tradition.

The *experimental* aspect of the new world-view implied a belief that man could not get satisfactory answers to the questions he asked about the nature of the world by reflection and discussion alone, but only by combining these with activities of a practical kind. Aristocratic disdain for manual work was now felt to be out of place, and many of the seventeenth-century scientists were keen to master the knowledge of craftsmen – even Newton ground his own lenses. If discoveries with practical applications were to be made, it was essential to learn as much as possible through direct physical contact with a wide variety of substances. In addition, the doctrines of the empirical philosopher Locke made sense experience a key feature in the acquisition of knowledge.

The Royal Society experimented with practical projects, often with little success; but Savery's steam engine dates from 1699 and it is worth remembering that in the development of steam power for a very long time theory was preceded by practice. A great deal of patient, practical laboratory work was undertaken in alchemy even by men like Boyle and Newton – work which produced few results at the time, but which paved the way for chemistry. Also, it may be remarked that mathematics and experimentation could be reconciled by emphasising the importance of precision measurement in experiments.

Finally we come to what some consider the most important characteristic of the scientific world-view: its *totalitarian* aspect. By the use of this term it is meant to suggest that the new methods and new knowledge were such as could be shared and applied increasingly by an ever-growing army of researchers, now efficiently organised in learned societies. More significant still, however, these methods and this knowledge could be applied to an ever-widening field. The greatest triumphs had been achieved in cosmology and mechanics, but what was to stop the conquest of other areas of study, of the inanimate world, then of the world of living things, even in the last resort of religion itself? For already there had been a deep invasion of the sacred, and a process of desanctification and disenchantment. In the world of 1700 the spirits and angels had been banished; for, in spite of the 'vast empty spaces' that terrified Pascal, there was nowhere to put them. In medieval cosmology though man's position was low in the sense of being far from the celestial spheres, it was still at the centre. But he was displaced from this centre by Copernicus, and his physical insignificance was to be repeatedly re-emphasised in succeeding centuries, for, as astronomy grew, so did the size of the universe postulated by astronomers. If God still remained, it was as a Great Geometer or Celestial Mechanic, not as a Jealous Lord to whom prayers should be offered, nor as a Loving Father in whose image man had been created. The men of science were devout, and proclaimed that they were motivated in their researches by the desire to glorify Him. Some time was yet to pass before atheism was to be accepted as the natural bed-fellow of science, though some scientists still manage to reconcile a scientific with a religious outlook even today.

The implications of this quantitative, analytical, experimental outlook, with its colonising and totalitarian propensities, were not all apparent, of course, for many years. Among those unacquainted with science, arguments still raged in 1700 (as they had in 1600) about whether or not men of that period were superior in knowledge to the ancient Greeks and Romans. But there is little doubt about the ultimate goal of the early scientists: there is much evidence in their writings that a desire for *control*, for power over nature, was their central driving force, and they believed that this power could only come through scientific understanding. Their methods generated knowlege which, whatever its limitations, led to results that were reproducible, repeatable and predictable, and thus ideally adapted for use in the control of nature. If research was undertaken for the greater glory of God, it was in the hope that practical benefits would also follow from it. The material world, far from being a barrier between God and man to be transcended rather than tampered with, was there to be investigated and revealed as a manifestation of God's infinite wisdom, and to be exploited for man's welfare and comfort. Thus the theme of power and control as the goal of science is an early one, often explicitly

spelled out by the early scientists. It is central to scientific endeavour. It later became fashionable in England to consider pure
research as intrinsically superior to applied science, and to deny
interest in the use of one's discoveries. These attitudes, however,
may well be considered as rationalisations; it remains likely that
control as a goal, though sometimes a hidden, implicit one, basically
motivates even the modern pure scientist.

Since the seventeenth century, man's capacities for exercising
control have enormously increased. He has extended his power
from small, local areas to the whole surface of the earth and even
beyond; he can harness the forces of inanimate nature on a huge
scale for the release of vast quantities of energy. His hold on the
biological world of plants, animals and other living organisms has
tightened; operations and activities which previous eras might well
have described as miraculous, are now routine, commonplace.

With relation to control, one set of concepts within science has
proved particularly fruitful. These are the concepts relating to
systems, and especially to the combination and working together of
the various elements of systems, particularly those which are automatic or self-regulating. Watt's governor in his steam engine is a
practical forerunner of the ideas of negative feedback in electronics
and of cybernetics in general; concepts of homeostatic equilibrium
in biological cells and of ecological balance belong to the same
family. Some structuralist-functionalist theories of society in sociology and neo-classical economic theory use similar concepts of
equilibrium. We may say that the modern scientific world-view
recognises intermediate levels of organisation, hierarchies of at
least partly autonomous systems and sub-systems, each the preserve of a different scientific discipline. It is not that the analytic
aspects of science have been weakened; the project of science still
seems to be ultimately to reduce all phenomena to manifestations of
'matter in motion' or, in more modern language, particles and fields
interacting; and the belief of science still seems to be that ultimate
understanding of this sort will confer omnipotence.

3   Science and Technology as Means of Control

The subject-matter of Part II provides a wealth of illustrations of the
scientific world-view, in its totalitarian, mathematical, analytical

and experimental aspects, applied to a wide range of activities with the aim of enhancing man's power to control not only his physical environment but also the behaviour of his fellow men. The very breadth of this range of activities, and the invasion of more and more fields (education, sport, work, sex, reproduction) that were formerly considered private or inappropriate for scientific analysis, testify to the totalitarian propensities of science. Mathematics – measurement, statistics and quantification in a variety of forms from abstract fundamental theory to simple accountancy – enters into every major problem field; the ubiquitous computer, that almost perfect example of a scientific product, reinforces the point. Analytic approaches to perceived problems, as systems analysis, operational research or in other forms, are so common and seem so natural to many people as to rule out other modes of attack. The experimental method, writ large, underlies much technical innovation, which, at the expense of tradition, embraces a restless and relentless commitment to progress and change.

We have argued that the motive force behind the scientific worldview is a desire to control nature, to master powers that can be used to make man less vulnerable to scarcity, natural hazards and disasters, and effectively increase his speed, mobility and mechanical strength. Already by the end of the nineteenth century steam, electricity and the internal combustion engine had given western man a sense of power and a degree of control greater than he had ever experienced before. Thinking of nature as an empire to be conquered, we find it extremely easy to extend this list of victories to the present day – high food yields; birth control and longer life expectancy; the harnessing of nuclear energy; telecommunications that are instantaneous and world-wide; the production of new materials and consumer goods in a constant stream – to select just a few examples considered in Chapters 6-11.

In human communities, however, control over people is at least as important as control over nature; indeed the two types of power are closely linked. In primitive societies belief in access to the magical secrets of nature by a priest class bestowed power on that class. In modern societies, science and technology do not develop in a political vacuum. The directions in which science spreads, and the forms taken by technical innovations, are determined by the political context – by such factors as the distribution of wealth and resources, the tenacity of ruling groups, and the relative strengths of various institutions and factions. Thus, as we have seen in Chapter 6, it can be argued that the 'Green Revolution' in agriculture in the Third World took place in such a way as to strengthen the interests of the large multinational companies who dominate the world markets in seeds, fertilisers, pesticides and food products, and to weaken the position of the poorest farmers. New technologies of production in the Industrial Revolution strongly favoured the rich factory owner at the expense of the poor labourer. Likewise, new tendencies in systems of transport, communications and mass media; shifting policies for energy, health, defence; strategies for research and development of all types, reflect the power struggles within society at the present day. Some people would maintain that it is not just that different groups have different priorities in applying already given technologies; rather, in societies which were politically fundamentally different from our present society, technology would develop in quite different forms, and even pure science would look different too. These are questions we return to in the next chapter.

Control over people is exercised through science and technology in many different modes which overlap and interact. First, as we have already suggested, in so far as technology is a key factor in industries of all sorts which play a major role in the economy, then it is involved in the whole politico-economic process which allocates goods and wealth in varying quantities to different individuals and groups. The distribution of the benefits of technology is an aspect of political control in the widest sense. These political dimensions are further discussed in the next chapter. But there are other, perhaps more immediately obvious ways in which science and technology can be used to manipulate people, forming a spectrum which includes at one end direct physical interventions by military and police authorities, and at the other the subtle abuse of scientific concepts and theories in political arguments where they are not necessarily appropriate or valid (e.g. see Chapter 7 Section 4).

We have written at length in Chapter 9 about the technology of war in the sense of open armed struggle between nations, but only mentioned in passing the question of military and police control over civilian populations. Since the Second World War a huge array of new technologies has been developed to facilitate surveillance, crowd control, interrogation of suspects and 'treatment' of prisoners. As well as employing older police methods, these technologies of repression draw very heavily on new developments in electronics and on medicine, pharmacology and biochemistry. Computer networks provide easy access to stores of information about very large numbers of people, information which can be collected secretly and in great quantity by phone-tapping, bugging and similar techniques. For crowd control there is a choice between impact weapons (water cannon, rubber and plastic bullets), chemical weapons (tear gas, CS gas and other irritants, hallucinatory agents), and many other exotic species (sound curdlers, epilepsy-inducing flashing lights, and so on). Suspects and prisoners, whether criminals, political subversives or other social deviants, can be injected with chemicals to tranquillise, harass, stupefy, nauseate or sexually neutralise them; they can be operated on surgically; or they can be subjected to systematic torture with a hundred different refinements (Ackroyd *et al.*, 1977; Schrag, 1980; Aubrey, 1981; Campbell, 1981).

Science has contributed much to these methods of physical restraint, persuasion and punishment, but it has equally transformed the more subtle modes of social control that operate through mass communications. Radio and television, with centralised programming and vetting of material broadcast, are ideal media for disseminating propaganda both political and commercial, and for moulding opinion. These technologies have for the most part been developed restrictively in such a way as to enable a small active group of broadcasters to deliver their message to a very large and necessarily passive audience, lulled or distracted by massive doses of sport and undemanding entertainment (see Chapter 10).

We have seen in Chapter 11 how the world of work has been subjected to scientific methods, by experimenting with new techniques, and by analysing and quantifying working activities. These approaches were first applied to manual tasks but are now increasingly evident in work involving mental labour; they make certain manual and mental skills obsolete. New microelectronic technologies, both in factories and offices, make possible the replacement of workers by machines and provide the means for very close supervision and monitoring of those who remain. The result may be large-scale unemployment, decreasing freedom and flexibility for the employed worker and increasing control of management.

Controlling effects on our consciousness, on our perceptions of ourselves and of society, are no less important than the more direct physical influences. At the level of these ideas and perceptions, even if we are sceptical, we cannot fail today to absorb the scientific ethos, just as no one in earlier periods could have insulated themselves from the religious ethos which science has now largely displaced. Science constrains us to think in certain modes. The scientific world-view begins to appear 'natural', and we forget it is a human, social construction, the result of particular conjunctions of events at a particular period in history. Assembly-line modes of production and bureaucratic social organisation, for example, are not absolutes like illness and death, forced on us by the nature of things and from which there is no escape; they are ways of doing things that science and technology have brought into being, but because of their pervasiveness, they deeply affect us and exert control over our conscious and unconscious thought. Science cannot provide answers to moral and ethical questions, though it may colour our thinking about them. We may fall into scientism, trying to use science dishonestly or wrongly to legitimate our political beliefs and practices. (The work of Darwin, for example, has been frequently invoked as a proof for the social philosophies of both the Right and the Left.) Finally, it should be said that many scientists seem unaware of these ideological aspects of science as control; they are usually only discussed by critics of science to whom we shall return in the next section. (Ellul, 1965; Marcuse, 1968; Berger *et al.*, 1977).

4 Benefits and Costs of Technology

The control science and technology gives in manifold ways over the external world and over people, entails obvious benefits and equally obvious costs. There is far from general agreement about the extent, the relative balance and even the nature of these benefits and costs. We shall look here at three different attitudes to technology, two extremes (enthusiastic, uncritical acceptance and hostile rejection), and one, to which most people subscribe in some form, occupying the middle ground. The latter view is that we must develop science and technology to serve us, but bear in mind that definite costs are entailed, the whole process needing to be subjected to careful control.

Each of these attitudes has a long history. Visions of industrial society based on technology are found in the work of Saint-Simon and Comte in the early nineteenth century, and were developed in very different directions by some writers and attacked strongly by others. The technocratic point of view in its simplest form is that the benefits of technology far outweigh the disadvantages that are incidentally and unintentionally generated, difficulties which in any case can be solved by using more technology. There is a technological 'fix' for all problems, it is said, whether they relate to food, energy supply, resources, pollution, or even to social evils like crime, unemployment and poverty; and there are technicians to provide solutions, social engineers as well as electrical and mechanical ones. For the technocrat, the coming to fruition of science and technology represents a great breakthrough in the affairs of man, not only because of the overwhelming benefits they bring, but because they also suggest to him rational methods of decision-making and government that increasingly render old political ideas obsolete (See Chapter 4). Political problems are really only technical problems, it is maintained, or are easily reducible to technicalities, soluble by 'experts' trained in science and therefore conditioned to think objectively. Though it may provide itself with a veneer of respectability through puppet parliaments or similar bodies, technocracy is profoundly undemocratic. Its style of rule may be more or less benevolently tolerant or brutally repressive, as writers like Huxley and Orwell have imaginatively suggested.

Elements of technocracy can be detected without difficulty in most modern governments, and in the EEC Commission, for example, and there have been interesting discussions about where real power lies, and should lie, in the modern state. The enthusiasm most people feel for the benefits of technology, however, falls far short of advocating technocracy, and it is considerably dampened by their increasing acknowledgment of environmental, social and other costs. To our list of examples of man's increased control over nature that has made possible unprecedentedly high standards of comfort must be added, as we have seen in Part II, a list of undesirable and unintended or unavoidable side-effects: environmental damage (pollution, deforestation, extinction of plant and animal species); ethical problems arising from modern medicine (transplant surgery, life-support systems, genetic engineering, increasing world population and increasing numbers of old people);

production of dangerous substances (pesticides, drugs such as thalidomide, high-level radioactive waste); accidents (in industrial plants and transport systems); exhaustion of resources; and so on. Casting a shadow over all modern life is the threat of war with nuclear, chemical and biological weapons, the possession of which in ever larger quantities tends to undermine rather than increase national security. Technologically primitive societies fail to survive or are destroyed in the modern world (Brain, 1972), and this has led towards a standardisation of cultures intensified by telecommunications networks that reach every remote region of the globe. And we have yet to include in the balance sheet those methods of control over people, outlined in the last section, which raise the possibility of future states far more repressive than the worst tyrannies that have so far existed.

The middle view, however, is that in spite of this depressing catalogue of drawbacks and dangers, we must press on with science and technology, committing resources on a large scale to research and development, and industrial innovation, with the aim of promoting economic growth. Most of the industrialised nations adopt this course of action though some with a greater level of commitment than others. The belief is that the threats and difficulties are containable, and the risks small, compared to the risks of being left behind in the international economic race.

Opposed to even the cautious promotion of science and technology is a set of more extreme opinions which question many of the alleged benefits of technology and raise objections to science at a very fundamental level. It is argued that we are enslaved by technology in its present form, in the sense that individual autonomy is undermined. Illich (1975), for example, suggests that we rely on others to a far greater extent than formerly for our food, shelter, work and health care. Our education is designed to adapt us to the procrustean bed of the modern industrial system; pseudo-needs are generated, while 'real' human needs go unsatisfied. We become hooked on high technology as on heroin; the technological fix is not only a technical solution to a problem, but the addict's shot in the arm that brings short-term relief, but eventually destroys. We come to rely more and more in everyday life on a variety of magic 'black boxes' whose functioning we dimly, if at all, comprehend. Is the ultimate aim, it is asked, to create totally controlled environments for us to live in, leaving us individually as helpless and dependent as a baby in the womb?

Currents of anti-scientific thought that oppose the scientific world-view in its totality, date at least as far back as the Romantic Movement of the late eighteenth century. Since then, poets and artists with rare exceptions, consciously in revolt against the soul-lessness of industrial society have, implicitly or explicitly, rejected the scientific picture. As against an analytical view of the world, they have stressed their sense of wholeness or oneness with nature, their intuitions of the interconnectedness of things ('We murder to dissect'). Against the quantitative aspects of science they set the impossibility of measuring love, joy, desire, happiness, grief, beauty – qualities they see as far transcending scientific truth. With their sense of tradition, and in their fear of natural forces that they experience as mysterious, imperfectly understood, they oppose the experimental, sorcerer's apprentice attitude. They feel threatened by the undiminishing rate of expansion of science. Above all, they suspect just what scientists consider their major strength – the attempt at objective assessment of predictable and repeatable phenomena, rather than the search for unique, highly subjective experience. With this denial of emotion, of human feeling, say the modern Romantics, it is no wonder that science is so destructive: the philosophical gap is a wide one, not easily bridged (Roszak, 1971).

It should be noted that current criticism of science and technology does not only come from politically radical Romantics, or poets and artists, but also from religious groups, such as some Islamic inter-preters of the Koran, or those American Fundamentalists, who accept literally the account of the creation given in Genesis. Cer-tainly, there is a great difference between the religious or even the humanist model on the one hand, and the out-and-out scientific-reductionist model of man on the other – 'twenty dollars' worth of chemicals', as someone has called it in a memorable phrase – a man of 'matter in motion' and nothing else, dehumanised, 'beyond freedom and dignity', according to the psychologist Skinner (1973). Scholars like Galileo, with their emphasis on primary qualities, had already drained the world of colours, sounds and smells, for these being secondary qualities had no existence except in the eyes, ears and nose of the spectator; in reality there were only grey, silent atoms, moving and interacting according to mathematical laws. Modern scientists, say their critics, would appear to have an even

bleaker view of the world than their seventeenth-century counterparts.

References

Ackroyd, C. *et al.* (1977), *The Technology of Political Control*
(Harmondsworth: Penguin).
Aubrey, C. (1981), *Who's Watching You?* (Harmondsworth:
Penguin).
Berger, P.L. *et al.* (1977), *The Homeless Mind* (Harmondsworth:
Penguin).
Brain, R. (1972), *Into the Primitive Environment: Survival on the
Edge of our Civilization* (London: Philip).
Campbell, D. (1981), *Big Brother is listening: Phone Tappers and
the Security State*, (London: New Statesman).
Ellul, J. (1965), *The Technological Society* (London: Cape).
Illich, I. (1975), *Tools for Conviviality* (London: Fontana).
Lovejoy, A.O. (1965), *The Great Chain of Being* (New York:
Harper Torchbooks).
Marcuse, H. (1968), *One-Dimensional Man* (London: Sphere).
Roszak, T. (1971), *The Making of a Counter Culture: Reflections on
the Technocratic Society and its Youthful Opposition* (London:
Faber).
Schrag, P. (1980), *Mind Control* (London: Boyars).
Skinner, B.F. (1973), *Beyond Freedom and Dignity* (Harmonds-
worth: Penguin).

13 The Control of Science and Technology

CHARLES BOYLE

1 Introduction

In this chapter we turn to the urgent question of how we can harness science and technology for our benefit, and at the same time minimise the associated dangers. Section 2 outlines different political perspectives; Section 3 deals with attempts at both national and international level to regulate our present technology; and Section 4 looks at the planning and assessment of technologies of the future. These three sections assume technology developing in more or less its present form; alternative approaches, hostile to large-scale, high technology are considered in Section 5. The final section offers a brief summary of different views of the future.

2 Political Viewpoints

In chapter 12, we drew attention to the enormous powers bestowed by science and technology. The question of who controls this power, by what means, and for what purpose, is thus a central, *political* question. Scientists, and especially teachers of science, are often very reluctant to acknowledge that politics has anything to do with science, but we have underlined the importance of political factors in Chapter 4 and in the discussions of the various topics of Part II. Here, with a slightly different emphasis, and as a supplement to these earlier analyses, we look at some ways in which, even when there is a broad consensus favourable to science and technology, there can be very different views about how they should be applied and controlled.

Questions about the control of science and technology are not simply questions of technical decisions to be made and practical measures to be taken, (e.g. on hazards of different sorts), they involve much wider issues – the extent of state control of industry and of funding of research, public participation in science policy-making, the safeguarding of civil liberties, and so on. One's understanding of the workings of society – whether, in the terms discussed in Chapter 4, one takes a consensus or a conflict view – will influence one's recommendations of the directions science and technology as key generators of wealth should take, and the controls to be imposed on them.

Let us consider, for simplicity, only three positions, at the Right, Left and Centre of the political spectrum: the conservative view, with its emphasis on the importance of market forces and of competition in ensuring efficient use of resources; the radical socialist view, showing concern about unequal distribution of wealth, and advocating public control of production, distribution and exchange; and the reformist (or, to the Marxist, revisionist) position, advocating the mixed economy, that is, one with some government intervention within an overall capitalist framework.

What are the main differences in the science and technology policies of Left, Right and Centre? The question is difficult to answer directly and in general terms; answers must be given in relation to the broad industrial, social and economic objectives of the different political groupings, or with reference to specific situations and examples. Science policy often does not exist as a separate scheme, but is imbedded in wider plans for the modernisation or reorganisation of industry, for example. Parties of the Left are likely to be concerned about wide distribution of both the benefits and costs of technology. In general they will favour higher public spending on welfare, tighter public control over high-technology industries, and research and development programmes carefully planned and directed towards public and national needs as they perceive them; they will consider high unemployment unacceptable. In the western world, the Left and Centre tend to be somewhat more sympathetic to public participation in planning, pollution controls, civil liberties issues and alternative technology (see Section 3 below) than parties of the Right. The latter, keen to boost company profits, to sell nationalised industries, to relax controls and encourage private industry in general, believe government influence should be limited to fields of military and strategic importance, or high-technology industries such as nuclear power, where investment risks are considered too great by private investors. In these areas, however, the contract system, particularly in the US, makes government money available in large quantities to private corporations. In practice, under the pressures of government, political ideals can be jettisoned in particular situations if it is judged to be expedient. In the UK, for example, a Conservative government nationalised a large private company, Rolls-Royce, in 1971; while a Labour administration in 1975 paid out some \$160 million to the American corporation, Chrysler, to keep their British subsidiary in business, rather than take it under state control (Open University, 1978).

We can illustrate sharp differences in political perspectives by looking at attitudes to the activities of multinational corporations (MNCs) in the problem areas examined in Part II. These large companies, operating in many countries, were important before the Second World War in the extraction industries, but have since become very powerful in the manufacturing sector as well. Committed to high technology, they develop it in their laboratories and apply it in the processes they use and in the products they sell – chemicals, cars, food products, weapons, pharmaceuticals, electronic equipment, aircraft, and so on. They could not function without world-wide computer, transport, telecommunications and information-collecting networks. The bigger MNCs have sales figures that dwarf the GNPs of many Third World countries, and even exceed those of some small industrial nations; the economic and political power they wield is therefore enormous, as is their influence on technology.

To many conservatives, MNCs represent a strong and welcome force for Third World development, as repositories of technical, managerial and commercial knowledge and skills, without which progress cannot be made by an industrialising nation. To many socialists they are anathema, spreading their tentacles in a global system of capitalist domination and exploitation, using bribes to keep puppet military élites in power, siphoning off profits in a very unequal system of exchanges, and reducing already poor countries to a state of abject poverty and dependence. The reformist, rejecting violent revolution as a permissible mode for eliminating gross inequalities of wealth, recognises the modernising and productive aspects of MNCs, while urging a strict control on their less salubrious practices.

The rich world (including the Soviet bloc) has only a quarter of the world population of 4 billion, but it has 90 per cent of the world's manufacturing industry, and spends 95 per cent of the total funds devoted to research and development. It dominates trade and patent systems, and hence technology; the Third World thus participates in these activities on very unequal terms, but it makes relatively small demands on total resources at present. It is estimated that the rich world's annual 'crop' of 16 million babies has four times the impact on world resources of the 109 million babies born annually in the Third World, so great is the disparity in consumption per head. In the industrialised nations life expectancy is 70 years, in the less developed countries 50. In the latter, one-fifth of the population suffers from hunger or malnutrition, and in the poorest countries 1 child out of every 5 dies before the age of 5. Estimated Third World unemployment rates are 30 per cent or higher (*Brandt Report,* 1980).

Science and technology, it is often argued, can make a significant contribution towards helping the world's poor, but the results to date are disappointing, as we have seen in Chapter 6, in the case of food and agriculture. The economic gap between the rich and the poor nations is widening, not closing. The reasons given will, of course, vary with the political philosophy of the speaker, as will the proposed solutions.

3   The Regulation of Technology

Control is a word that has many shades of meaning; here we shall distinguish two main aspects of the control of science and technology: the regulation of what is in operation already (discussed below); and the planning of new technologies for the future in line with visions of what the structure and goals of society should be (discussed in section 4).

Regulation implies adjustments to a machine or system functioning fairly well. It is concerned with things which go wrong or are unforeseen, with reacting rather than initiating. But the image of technology as a machine to which delicate mechanisms of control can be applied, for example to speed it up or slow it down, is not a particularly good one; the regulation of technology at present is often largely a matter of responding retrospectively to unexpected side-effects and dangers, rather than of making fine-tuning adjustments in anticipation of a predictable difficulty. Arrangements for adequate regulation lag well behind technological development, and too often are brought into operation only as a result of accident, malpractice or serious damage of one kind or another to people or the environment.

The hazards with which regulation has to cope fall into many different categories; how one groups them and ranks them in order of importance is to some extent a matter of personal prejudice. One approach is to think of dangers arising from intentionally destructive, harmful or repressive technologies, and distinguish these from the unintended hazards of otherwise benevolent technologies.

With reference to the first group, in Chapter 9 we identified arms control as one of the most pressing and intractable of modern problems – work on nuclear, chemical, biological and other weapons continues to expand to such an extent as to suggest to some people that in this area technology is out of control. Increasing concern is expressed also about the power placed in the hands of the police and others by the rapidly developing range of what might be called 'Big Brother' technologies referred to in Chapter 10. The question here is how to ensure adequate means for the effective policing of society without undermining individual rights to privacy, freedom of thought and communication.

Our second category of hazards includes those arising from various technological products and processes, most but not all the result of recent developments, designed for particular purposes, but also giving rise, sometimes unpredictably and in the short- or long-term, in the home, workplace or general environment, to undesirable consequences. A catalogue of such hazards would be extremely long. Certain emotive words – thalidomide, Flixborough, Aberfan, Ronan Point, Torrey Canyon – associated with disasters, make headlines and linger at least for a short time in the public consciousness. But this list is meaningful only perhaps to English people over a certain age; the Italians have their Sevesos, the Japanese their Minimatas, the Americans their Love Canals. These names, recalling explosions at chemical plants, contamination of the environment by toxins, drugs that turn out to be teratogenic (monster-producing) poisons, mechanical failures of man-made structures, pollution by oil-tanker accidents, and the like, must be supplemented by other less spectacular but more common, and in some cases even more deadly, occurrences – car and air crashes, diseases, arising from exposure to asbestos or radioactive materials or other carcinogenic and mutagenic agents.

Many of these tragic occurrences give rise to immediate official inquiries, some of which result in rapid legislation. In other cases the wheels of officialdom and justice turn very slowly, if at all, and little effective action may be taken. Victims may have great difficulties in getting compensation, and may even have to struggle for official recognition that their complaints are justified. Procedures and practices vary; some countries show greater concern than others.

By what means can technology be regulated to minimise its hazards and their possible effects? Who are the interested parties? In other words, who is to be controlled, by whom, and how? These are the questions to which we must now turn our attention.

The principles of law, in its various forms and acting through its various institutions, provide a primary means of controlling human behaviour. Legislation is, in most cases, the most effective way of regulating technology, particularly legislation exerting economic pressures on firms to bring about desired changes in technological practices, whether they are in the form of grants or subsidies to firms engaging, for example, in approved pollution-control practices, or financial sanctions against those who are not. The pressure of public opinion is also important in serving to initiate work on the introduction of new legislative or economic measures. In the last resort only widespread individual concern and individual actions can make controls effective.

3.1   International Regulation

The problems of the international regulation of technology are in many ways both the most difficult and the most urgent. There is no international body with sufficient authority to impose its laws on the various nations of the world and to enforce them, in the way that a modern nation-state can exact obedience to *its* laws from all but the most powerful of its citizens and organisations. Thus the world-wide regulation of technology depends to a great extent on international agreements entered into more or less freely, but capable of being ended by new governments or as a result of changed circumstances.

Many difficulties stand in the way of reaching technological agreements. A great and almost unbridgeable gulf of hostility and distrust separates the NATO and the Warsaw Pact alliances, and this has led to very little progress being made on agreements on arms control and military technologies and on world arms sales limitations. Similar dangerous gulfs exist between many other nations and groups of nations. The failure to control arms sales points up, too, the economic competition, fuelled and sustained by new technologies, that is another divisive feature of the modern world. The US, the EEC, Japan and other countries are unlikely to agree to international regulation of their technologies that would weaken their competitiveness on world markets, unless there are very clear non-economic gains to be achieved.

For many people, as we have already suggested, by far the greatest threat from technology comes from the vertical and horizontal proliferation of nuclear and other weapons, and from the almost complete failure by nations to date to reach significant agreements on ways of limiting or reversing this technology of destruction. On this view the single most urgent technological issue of the next decade is the international control of military technology. Because of the enormous financial resources made available for military purposes, and because of the intense intellectual efforts invested in devising weaponry and in inventing diverse ways of outwitting or misleading the enemy, the problem is one of great technical as well as great diplomatic intricacy. As we have indicated in Chapter 9, the failures of the last few years (among them the non-ratification by the US of the SALT II Treaty between the US and the USSR, and the lack of progress at the UN Second Special Session on Disarmament in 1982), are indeed dispiriting; and even the successes do not lead to much optimism because of their very limited nature. ('Agreements by both sides not to bolt their planes to the ground' is how one cynic has described them.) There has been a deterioration in Soviet-American relations, and the recent ending of various agreements on scientific and technological co-operation, that have been in existence since the early 1970s, contributes to a bad climate for further negotiations. On the other hand, a new round of talks is under way, – renamed START instead of SALT (Strategic Arms Reduction, rather than Limitation Talks) – but these may, with adjournments, continue for some years before much progress is made. The growth of peace movements in the West is a new feature, but whether these will retain their momentum and influence, or fade like the earlier CND, remains to be seen.

The export of nuclear power stations has led to the spread of nuclear technology and the establishment of uranium-enrichment or plutonium-separation facilities in more and more countries. The number of nuclear-weapons states is increasing, though of course not all nations capable of producing their own nuclear bombs choose to do so. The Non-Proliferation Treaty, which came into force in 1970, had 115 signatories by June 1981, but it was not signed by France, China and India (known to have nuclear weapons), nor by Pakistan, Israel, Egypt, Brazil, Argentina and South Africa (countries presumed either to have nuclear weapons or to be capable of making them within a short period). These countries argue that the super-powers have not honoured their commitment in the Treaty to reduce their stockpiles. The nations which are suppliers of nuclear technology impose on clients safeguards designed to prevent the diversion of nuclear equipment or materials to weapons production, and it is the duty of the International Atomic Energy Authority (IAEA) to report violations of these safeguards to the UN Security Council. Many of the technical problems have been studied in various International Nuclear Fuel Cyle Evaluation (INFCE) meetings which have provided a useful forum for the discussion of proliferation issues, but have shown that these are political, not simply technical. In the last resort, due to the relative powerlessness of the UN and the other international organisations, and due also to the increasing amounts of weapons-grade material being produced, there is little that can be done at present to stop a determined nation from manufacturing nuclear weapons if it wishes to do so (Gummett, 1981).

Few among even the most ardent advocates of deterrence theory feel that world security would be strengthened by more nations having their own nuclear deterrents. Outside the nuclear field, as we have seen in Chapter 9, sales of other deterrents – improved conventional weapons, some of which are as destructive as small nuclear devices – continue to increase. We may add to these the possibilities of new ranges of chemical and biological weapons, work on which continues against the spirit of present agreements, which however are weak. These developments amount to arguments for some sort of powerful international body to regulate military technology, but the difficulties in setting it up seem almost insuperable. It may only be formed, if at all, after a disaster, after nuclear weapons, perhaps, have been used either by accident or design in some part of the globe.

As an example of an unwanted side-effect of technology with clear international implications we may mention pollution, a subject of concern at many conferences during the 1970s, since it does not stop at national frontiers. Sulphur released from coal burned in the power stations of the UK and Eastern and Western Europe gives rise to 'acid rain' on Scandinavian countries, and elsewhere, killing aquatic life in lakes and rivers and seriously damaging forests. Likewise, discharges of waste into seas almost enclosed by land, such as the Baltic or the Mediterranean, cause pollution that affects many countries. In situations such as these the countries involved will often have different cultures and political systems; they may well be at different stages of industrial development and have different priorities; some of them may be mutually antagonistic. Even if the necessary technical measures are clearly understood and unanimously accepted, which is unlikely, a strenuous diplomatic effort over a long period is still required to get agreement on the funding of research, action to be taken in the short and long term, arrangements for monitoring, and so on.

Third World countries, in deliberate efforts to attract industry, often have much slacker pollution regulations than those of the developed world; and they and the multinational companies which operate in them resist international pressures to tighten these controls. In the developed countries, too, which as well as being the major consumers are also the major polluters, governments are often persuaded to set aside or delay for economic reasons, new, expensive pollution-control measures. (See Chapter 5, Section 3.) Concern about pollution must be placed in a context of wider concern about the environment as a whole, about the need, for example, for a greatly improved environment in terms of water, nutrition, sanitation and shelter for the world's poorest people. Organisations such as the UN Environment Programme (UNEP) suggest that the control of pollution, the preservation of endangered species and the protection of forests are all simply aspects of the one environmental problem of respecting and sharing the earth's resources, and ensuring minimal living standards for all. Here again many of the problems are those of political will rather than technical knowhow, and there is a need to consolidate and supplement the work of existing international organisations.

Much international technology transfer, as we have seen, in-

volves multinational companies since they hold most of the world's
patents, and it is not surprising that anxieties have developed about
the practices of these organisations. A full account of the regulation
of technology in its international aspects would pay detailed attention to these practices and to the arguments that have been
advanced, for example by the Brandt Commission, for controls
over them. Some obvious abuses, such as the sale in the Third
World of drugs, pesticides and other potentially dangerous substances, which are banned in developed countries, have received
fairly wide publicity. (See Chapter 6, Section 7). Less known,
perhaps, but economically far more important is a whole range of
doubtful or unethical political and commercial activities (support
for illegal regimes, illegal payments, certain intra-firm trading
practices, profit shifting for tax avoidance, etc.) which have led to
the establishment of the UN Centre for Transnational Corporations
as a watchdog. Trade unions, too, have been trying to forge international links to enable them more effectively to resist MNC
management pressures on labour forces. There have been calls for
the formulation of codes of conduct, for example, in technology
transfer, and for changes in patent laws. Such calls can be seen as
part of a much broader demand for basic reforms in world trade that
would produce a more equitable system. (*Brandt Report*, 1980).

3.2   National Regulation

If we leave aside military questions, the first and the most important
attempts to regulate technology occur at local and national level.
Governments and other bodies have a duty to protect the health and
well-being of citizens, and through legislative and economic
measures they have the means to discourage or prevent activities
leading to harmful consequences. Some of the issues which arise,
such as pollution or the activities of MNCs, are not to be confined
within national frontiers; they spill over, as it were, into the international arena. There is thus a certain amount of overlap between
this section and the last; and, since it is difficult to discuss entirely
separately the regulation of technology at present in use, and the
planning of technology for the future, there will be links with the
next section too.

   We again make the distinction between technologies deliberately
designed to damage, control physically and restrict, and those from
which harm arises incidentally as an unwanted side-effect. In the

first category are the new technologies available to police forces which, as we have noted in Chapter 12, have developed dramatically in recent years. No modern industrial nation, it would seem, can exist without police to enforce the law and maintain internal stability, backed up by other para-military and military units. In a democratic society, there is the belief that police powers must be clearly defined and limited, and abuses must be seen to be dealt with effectively, if individual rights to privacy and to freedom of belief and communication are not to be undermined. There has been a strengthening of police powers, both through new technology and new legislation, in many countries including the UK, as a response, some say, to increasing terrorism; others, to increasing unemployment. Legislation to control the abuse of new technologies inevitably lags behind the technical developments themselves, but the consequences of unregulated surveillance, information-gathering, and intimidation are so serious as to threaten the very foundations of civilised societies. Urgent and sustained government action is needed, especially in those countries which have been slow to recognise the dangers.

Related to questions of information-gathering are questions of access to official information. A contrast may be drawn between the relative openness of the US government and the secretive, closed nature of government in the UK. Americans can, with little difficulty, obtain the sort of facts and figures relating to the work of government departments (not connected with national security) to which the British are denied access by the Official Secrets Act – an Act, it is often suggested, designed rather to protect officials rather than secrets. This has serious implications for public participation in the regulation of technology, as it may prevent concerned groups from obtaining vital information needed to challenge misguided official policies or even to identify official mistakes. Attitudes towards public participation vary: some maintain that discussion of technological issues should be left largely to politicians, civil servants and technical experts; others argue that wide public debate is a prerequisite of democracy, and is needed in any case to ensure that sensible and effective decisions are taken; many are apathetic, feeling perhaps powerless to influence events, even if they understand the complex technicalities involved.

It is said that eternal vigilance is the price of liberty; maybe it is also the price of safe technology. Who are the watchdogs, the groups and bodies concerned in one way or another to ensure the safety of modern technology? We shall refer mainly to examples of the situation in the UK; while many of the British organisations have their counterparts elsewhere, there are significant differencies between countries both in attitude and practice, and it is difficult to make broad generalisations.

At the base level are all the individuals who make up society and who, in their workplaces or outside them, may be at risk from accidents and bad practice. Some of these individuals are also scientists and engineers with a particularly strong sense of personal responsibility about misapplications of their work or inadequate safeguards or some such matter, who may express their unease publicly and incur sanctions from colleagues and superiors. These, and other men and women with shared values, grievances and commitment, may join together in groups to draw attention to certain abuses of technology and to advocate reforms. Examples of such groups are the Friends of the Earth, the Socialist Environment and Resources Association, the British Society for Social Responsibility in Science, and many *ad hoc* committees formed locally to oppose a planned new development such as an airport, motorway or power station. Trade unions may have general interests in the control of technology in relation to unemployment, for example, but they are also specifically concerned with the health and safety hazards to which their members may be exposed at work The Health and Safety Commission, including the Alkali Inspectorate, the Factory Inspectorate, the Nuclear Installations Inspectorate and various others, is the official body with professional and statutory obligations to monitor the technological activities of industry and to ensure they are being carried out safely within the law.

At higher levels still are Royal Commissions and Advisory Committees of one type or another which are set up by governments, often after accidents and to appease public disquiet, to investigate, analyse and make recommendations. Thus in Britain the Dunlop Committee on the Safety of Drugs was set up after the thalidomide tragedy which led to the birth of many seriously deformed children; as a result of public anxiety the Royal Commission on Environmental Pollution has carried out various inquiries, and since 1971 has published a series of reports; the Robens Committee reported on Safety and Health at Work in 1972. At the apex of the power pyramid are parliament and the government who may act on the recommendations made to them and pass appropriate legislation (Open University, 1978).

Accounts of this legislation in the UK, and of the corresponding measures elsewhere, and discussions of important case studies in the safety of drugs, pollution, health and safety at work, as well as in many other technologically significant areas, are readily available, and the reader is referred to them for details. Here there is space only to comment on a few general points.

First, there is a clear need to separate sharply the planning and operation of a technological process from its regulation. An inspectorate will ideally have no links with the management of the industry it is required to inspect, nor will it develop a too generous or sympathetic understanding of the industry's problems, if effective regulation from the point of view of the public is to be obtained. The Alkali Inspectorate in the UK has been accused in the past of having had a much too complacent and accommodating approach to the control of air pollution. In the field of nuclear energy, to take another example, in view of the possibly horrific consequences of mistakes, it is essential that the inspectors called in to examine the safety of nuclear reactors should be independent. It may, of course, be difficult to find truly independent outside opinions in highly complex but narrowly specialised subjects such as nuclear reactor safety, where there are only small numbers of experts, all more or less well-known to one another, who may even have been colleagues in the past.

A second question concerns the type of legal mechanisms thought best for effective regulation. To what extent should recommendations and guidelines be spiced with legal compulsion? There are revealing differences between the UK and the US in their approaches, for example, to air pollution. Each approach has its advantages and drawbacks. The Americans lay down specific numerical emission limits, and an organisation with a factory from which the emissions are above these limits can be challenged in court. The British, favouring less adversarial and perhaps less aggressive methods, adopt the more elastic concept of 'best practicable means' which they require to be used by polluters to control their emissions. It is much more difficult for British than American citizens to take legal action against polluters.

Thirdly, rapid action must be possible when dangers are identified; indeed the need is for anticipation of possible harmful consequences arising from the use of new chemicals and processes. Too often, at present, and the thalidomide and asbestos cases illustrate the point, the first signs of something wrong are greeted with denial or disbelief, and it is only much later, when great damage has been done, that appropriate measures are taken. Capitalist economies have been very successful in promoting technological innovations. The pressure for profit, and the drive to expand, to capture new markets, lead to thousands of new chemicals being produced each year, some in large quantities. (It may well be, particularly in the pharmaceutical industry, that there is unnecessary duplication.) Since the financial rewards in marketing new products ahead of rivals are considerable, so also are the temptations to take risks and cut corners in testing for toxicity. In these circumstances strict regulation by government is essential; exhaustive testing of all new products may not be feasible, but if errors are to be made they should be on the side of caution.

Finally, it may be said that the effective regulation of technology depends ultimately not on special inspectorates and committees but on public attitudes. Where there is genuine, informed concern and a developed sense of social responsibility and public accountability, standards of safety will usually be high, though there is a surprising tolerance of very large numbers of deaths and injuries from road accidents. Perceptions of risk from various hazards by the general public are often at odds with those of risk assessment experts, whose quantitative approach is, however, open to question. Wider awareness and appreciation of the issues, promoted by educators and the media, might or might not close the gap between the public and the experts, but they would seem to be essential ingredients of safe technological practice. Danger thrives on apathy and ignorance.

4 Provision for the Future

The day-to-day problems of the regulation of technology are those which absorb most of the attention of governments whose time horizons usually extend only for a few years, as far as the next election. Equally, or even more important, however, is the long-term planning of technology in relation to economic, political and social objectives. Much of the UK research and development programme is drawn up, and carried out, in a fragmented, *ad hoc* manner; even in the single field of computers, it is a question of people 'doing their own thing'. Senior British scientists in the House of Lords and elsewhere have frequently argued for much more coordinated efforts, centrally organised, from a strategic viewpoint. The Japanese Ministry of International Trade and Industry (MITI) has been seen to be very successful over the past thirty years in framing ambitious policies, harnessing private and government resources in pursuit of them, and achieving results.

We have already noted that in manufacturing nations, whatever their political complexion, there are few governments who are not committed to technological innovation as a key element in maintaining their competitiveness and in achieving economic growth. The Wilson government of the late 1960s in the UK, and the Mitterand government of the early 1980s in France both laid particular emphasis on science and technology as generators of national prosperity. The belief is, as the British Cabinet Office Advisory Council for Applied Research and Development (1979) put it:

We have no option but to attempt to match the productivity and product quality of our overseas competitors, to concentrate our efforts on those industries where we have most chance of success, and to adopt as fast as possible, technical innovations from abroad, as well as those developed at home, which will enable us to do this.

It is to be remembered that some of the industrially most advanced nations are also committed to huge spending on military research and development. The unwanted side-effects of technology we have looked at are not taken as real threats, the general feeling among most politicians being that even if these effects became much more serious, they could still be coped with easily. One possible exception is the problem of technological unemployment, discussed in Chapter 11, with its threats to social stability, but even this has risen to high levels in some countries without many of the drastic repercussions that were predicted.

The decisions to be taken, then, are generally seen to be not about whether to proceed with technological innovation, and if so how fast, but rather about the choice of priorities to be pursued with full vigour. Past mistakes may be instructive: British officials can reflect ruefully on some expensive failures in the nuclear industry, and on the huge sums poured into the Concorde airliner and now irretrievably lost; Americans may be thankful they dropped their plans for supersonic transport, but may wonder if the Apollo project to put a man on the moon, though a boost to prestige, was really worth the resources consumed by it, and if these resources could not have been used, even in the field of space research, to much better effect.

The need is for effective evaluations of proposals for large technological projects, and also for mechanisms that can rapidly bring to a halt runaway schemes like Concorde, which, regardless of their lack of economic viability, once underway develop momentum as more and more people become committed to them. In evaluating proposals the techniques of cost-benefit analysis have long been applied by economists to provide background information for decision-making. It is clearly important to try to quantify both costs and benefits, but even when only narrow economic aspects are taken into account the difficulties and dangers are considerable (Freeman, 1974; Collingridge, 1980). The difficulties arise from the uncertainties of the future; too often current trends are simply extrapolated without much thought. The dangers arise from the temptation to oversimplify, to forget the uncertainties and margins of error (economics is regarded by many physical scientists as a one-digit or one-significant-figure 'science') and to emphasise quantifiable at the expense of non-quantifiable factors. Since the mid-1960s attempts have been made to broaden such studies by including environmental and social costs in the calculations. The scientific approach is to itemise and quantify these costs, reducing them to money terms. But what is the cash price of an historic church that has to be demolished in the course of a new development; or of lives that may as a result of increased accidents have to be sacrificed? To many just to ask such questions seems to betray crass insensitivity and lack of humanity.

Technology assessment need not, however, reduce everything simplistically to figures, in the mistaken belief that figures are somehow value-free. The idea of an independent but widely representative body of wise men and women studying new proposals carefully with all their social implications and possible repercussions, and making known their views to the public, is surely a good one. In the United States an Office of Technology Assessment was set up in 1972 to assist Congress and inform the public, and this latter function is partly fulfilled in the United Kingdom by Royal

Commissions.

Critics of technology assessment fall mainly into two groups – those who say it is not needed since present market mechanisms are adequate to ensure that socially desirable technology is developed; and those who agree with it in principle, but argue that it is ineffective in practice. The latter point to the difficulties in getting genuine public participation in the assessments, especially from the poorer sections of society; they feel that new technologies always leave the disadvantaged worse off than before, and that, in any case, in spite of their supposed independence, technology assessment organisations inevitably end up as tools of government. The more extreme critics see technology assessment as a cosmetic and therapeutic exercise on an unjust system which should not be propped up but replaced, if proper distribution of the fruits of science and technology is to be achieved (Elliott, 1976; Cross *et al.*, 1978).

How secure *are* our present systems of government in face of technological change? May it not be, as Ellul (1965) has suggested, that various technological changes, which taken separately seem to amount to little and to be easily accommodated, might still together add up to something much greater, just as several minor, localised slippages of snow can provoke an avalanche? Will a revolution in our attitudes occur? In the light of high unemployment and after a period of unparalleled success in technological innovation, is capitalism now threatened with 'terminal illness', or as in the past will it seize on new opportunities and, perhaps in some new disguise, regain its vigour? Is Russian communism too rigid to accommodate the changes new technology will thrust upon it? Will the Chinese variety prove more resilient? Or will some political philosophy, a strange blend of permissive Left and authoritarian Right, emerge on the back of technology to carry the day? Only the boldest prophet will venture to give unqualified answers to questions such as these.

We have already drawn attention to the urgent need, through agencies with appropriate powers of inspection, for international regulation of presently existing technology especially in the military field. Many of our remarks apply with even more strength to the control of future technology now at the planning stage, which may well prove a Pandora's box filled not only with new, cheaper and more deadly weapons, but also with new issues to quarrel over, and new threats to the sovereignty and independence of technologically backward nations. In the next decades, whether in the hands of private corporations, governments or international agencies, technology can be expected to attempt to colonise all those regions into which it has so far sent only advance raiding parties – Antarctica, the sea bed and outer space (not to mention the 'inner space' of the human psyche). In the exploitation of the oceans and of vast new mineral resources, and in intensified telecommunications through space, the scope for conflict is great, but so too perhaps are the rewards of reaching agreements.

To every attempt at balanced, reasonable analysis a final comment must be appended. However wise, powerful and rational the watchdog and planning organisations set up by governments may be, there will still remain in the general consciousness, like an iceberg mostly submerged, a realm of technological nightmare that draws on images from the past (Hiroshima, Dresden, the Nazi extermination camps), and on fears of the future. Devastated by nuclear war, populated by robots, clones, man-animal monster hybrids, with a handful of human survivors clinging desperately to a few remaining civilised values, the landscape of this nightmare has been described in science fiction in countless variations. The images are all the more powerful for being surrealist, irrational. Science has not banished the irrational; on the contrary, it has in many ways fertilised it and given it a new lease of life.

5   Alternative Technology

We turn now to an attempt to escape from the technological nightmare, to a radical critique of many of the ideas and approaches outlined above. Far from seeing technology, in its present form of large-scale 'hard' or 'high' technology, as a net benefit, a provider of high living standards, with the promise of even better things to come, certain radical thinkers emphasise what they see as its alienating, dehumanising effects, its subjection of workers to the requirements of machines, its basic destructiveness. Some place the main blame on what they see as our present, essentially exploitative, political systems, the capitalism of the West and the state capitalism of the East European bloc, as that variety of communism is sometimes called. Others, as we have seen, have deeper reservations about the whole scientific world-view. In any case, what is needed, it is argued, as resource shortage and environmental constraints force us to replace our present technology, is a new kind of less intimidating technology, that rejects much of the conventional wisdom of today's technocrats. Different adjectives have been used by different writers to label their versions of this new technology: radical, liberatory, convivial, soft, or simply (the term we shall use) alternative. Robin Clarke (in Dickson, 1974) has given a comprehensive list, with a pronounced utopian flavour, of the characteristics of this alternative soft technology, set alongside those of present hard technology, and these are reproduced in a somewhat amended form in Table 13.1.

To many contemporary officials the soft technology society, with its ethos of co-operation, respect for nature and self-sufficiency, seems a hopelessly idealistic project, with no chance of being realised in the foreseeable future. Certainly, in a fully developed pure form it would appear to be remote. Nevertheless to escape from some of our present predicaments and dangers, and to establish a sustainable economy, we shall have to move towards alternative technology in some fields. The choice is fairly clear: policies can be based on a continuing commitment to high technology, high energy solutions, or to a gentler, alternative technology approach that would be very selective in its choice of tools.

In Part II we hinted at some of the policy alternatives – agriculture could return to more labour-intensive, mixed farming methods, and away from agrichemical, agribusiness prescriptions; medicine could adopt holistic rather than scientific, drug-dominated treatments. Energy could be carefully conserved, and obtained from 'renewable' sources rather than nuclear power stations; defence planning could be oriented more towards individual citizen resistance and repelling invasion, and less towards threats of mass and self-destruction. Communications networks could be decentralised and made more accessible and open. Work in industry could be organised to be more creative and less repetitive and soul destroying. In transport, building, education and use of the environment, to mention other areas we have not discussed, there are similar choices to be made. Significant changes in the mix of hard and soft technology will not occur overnight, and indeed may not occur at all, unless triggered by some large-scale catastrophe that would bring about fundamental reappraisal. We may

Table 13.1: Clarke's characteristics of (a) hard and (b) soft technology societies.

	(a)	(b)
ecologically	unsound	sound
energy input	large	small
pollution rate	high	low
materials, energy sources	non-renewable	renewable
functional or	limited time	all time
mode of production	mass production	craft industry
specialisation	high	low
social organisation	nuclear family/city	commune/village
attitude to nature	alienation	integration
politics	consensus	democratic
technical boundaries set	by wealth	by nature
trade	world-wide	local bartering
local culture	destructive of	compatible with
technology misuse	likely	safeguards
attitude to other species	destructive	supportive
innovation regulated by	profit and war	need
economy	growth-oriented, capital-intensive	steady-state, labour-intensive
young and old	alienated	integrated
organisation	centralist	decentralist
efficiency increases with	size	smallness
operating modes	very complicated	understandable
technological accidents	frequent, serious	few, unimportant
solutions to problems	singular	diverse
agricultural emphasis	monoculture	diversity
criteria highly valued	quantity	quality
food production	specialised industry	shared by all
work done mainly for	income	satisfaction
small units	totally dependent	self-sufficient
technology and culture	alienated	integrated
science performed by	specialist élites	all
work/leisure distinction	strong	weak
unemployment	high	none

Source: Dickson, 1974.

think of potential pathways of development, open for the moment, not yet closed off by circumstances.

There is one major area in which alternative technology would seem to have very distinct advantages – the Third World, most of whose countries have large unemployed labour forces and little financial capital, technical knowledge or managerial experience, so that their need is for simple labour-intensive rather than capital-intensive industries. As many writers have pointed out, much of the technology transferred by the West to the Third World has been developed by the West for its own purposes, and is quite inappropriate in scale and complexity for developing countries, increasing their debts and their foreign dependence. Schumacher (1973) in *Small is Beautiful* challenged accepted ideas about economies of scale (e.g. that bigger units of production always lead to cheaper products), and argued for the introduction of intermediate technology – intermediate, that is, in cost and size, between primitive tools and modern technology, between the spade and the combine harvester. Illich (1975) has written of 'tools for conviviality', which would enable the world's poor to take control over their lives and to preserve their dignity, identity and independence.

China is an example of a poor country with a strong spirit of independence and self-reliance, and a strong drive towards social equality, which has sought, through its policy of 'walking on two legs' to develop and apply appropriate technology in its social objectives. Just as walking entails the coordination of two different limbs, so successful social and economic development requires the harmonious co-operation of contrasting pairs of sectors – agriculture and industry, large and small enterprises, hard and soft technology. Recently, however, China has been leaning more heavily on imported high technology.

Not only in the Third World, but in the science and technology policies of all countries, there are tensions and inherent contradictions, benefits and (sometimes hidden) costs. For each country what is needed is an appropriate balance, a judicious mixture of hard and soft elements; what the relative weightings of these in the mixture should be is, however, a very open question.

6 Conclusion

Futuristic novels and projections of the future written in the past often have a rather quaint flavour, with one or two surprisingly accurate predictions, but very wide of the mark about many developments we now think of as natural and obvious. There is no reason to believe that the insights of present-day writings will prove any better, though in quantity they exceed those of the past. Science fiction has for some time been a popular and well-established genre, and many volumes of future studies have been written in the last decades, and are still being written, reflecting the anxieties of a nuclear age.

Of these studies it may be said that there seems to be no single aspect of the future about which they are all agreed. Some of the authors are optimistic about developments which are just those which make other authors despair. Among the pessimists there is a divergence of opinion about which threat is most likely and most to be feared – resource depletion, degradation of the environment, collapse of the world monetary system, or nuclear war.

We have suggested that to understand in all their diversity the philosophies, doctrines and ideas of today that are influencing the world of tomorrow, it is necessary to consider a full range of perspectives, and attitudes varying from distrust or hostility to enthusiastic acceptance, and from left to right in the political spectrum. We may think in terms of a diagram (see Figure 13.1) with a horizontal and a vertical axis representing, respectively, political orientation and response to science and technology. Technocrats, of both West and East, who believe in technological expansion and the technological 'fix', and in sustained economic growth as a consequence, will belong far up in the top two quadrants, to the left or right depending on the extent of their commitment to greater social equality and on the means to be used to achieve it. Conservative traditionalists and religious fundamentalists fit in the lower-right quadrant, while in the lower-left corner permissive Marxist and other left-wing advocates of alternative technology jostle for space. The great majority of people at present belongs to the top quadrants, not too far up the technocratic scale perhaps, but the numbers below the line could grow, especially if, as a result of accidents or dashed expectations or for other reasons, disillusionment with science and technology increases.

To many observers, by no means all on the political Left, it would seem, too, that without a definite trend towards greater economic equality between rich and poor nations, explosive pressures will build up that could be disastrous.

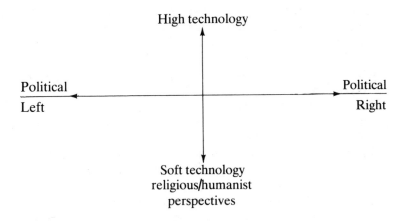

Figure 13.1: Political and technocratic attitudes

Each new generation feels itself faced with crucial tasks, but the generation in power over the next decades will have a particularly heavy responsibility, in that the resources it allocates, and the decisions it takes, will, to a greater extent that in the past, have irreversible effects, cutting the world off from certain paths of development and guiding it to others. The possibilities are there for almost total destruction, or for enslavement in an inhuman, world-wide, police state. On the other hand, we have the potential to abolish scarcity and bring slowly into being societies, by no means one-dimensional and homogeneous, but in which every individual would be provided with the basic material necessities, and could find a large measure of joy, happiness and fulfilment.

References

Advisory Council for Applied Research and Development (1979), *Technological Change: Threats and Opportunities for the UK* (London: HMSO).

Brandt Report (1980), *North–South: A Programme for Survival*, the Report of the Independent Commission on International Development Issues, (London: Pan).

Collingridge, D. (1980), *The Social Control of Technology* (London: Pinter).

Cross, N. *et al.* (1978), *Man Made Futures: Readings in Society, Technology and Design* (London: Hutchinson).

Dickson, D. (1974), *Alternative Technology: The Politics of Technical Change* (London: Fontana).

Elliott, D. and R. (1976), *The Control of Technology* (London: Wykeham).

Ellul, J. (1965), *The Technological Society* (London: Cape).

Freeman, C. (1974), *The Economics of Industrial Innovation* (Harmondsworth: Penguin).

Gummett, P. (1981), 'From NPT to INFCE: developments in thinking about nuclear non-proliferation', *International Affairs*, Autumn.

Illich, I. (1975), *Tools for Conviviality* (London: Fontana).

Open University (1978), Course T361, *Control of Technology*, Units 1, 3–4, (Milton Keynes: Open University Press).

Schumacher, E.F. (1973), *Small is Beautiful: A Study of Economics as if People Mattered* (London: Blond & Briggs).

A Guide to Further Reading

Our aim here is to indicate literature sources for the reader who wishes to explore in greater depth topics discussed or outlined in the main text. We provide a short list of books which we believe will be interesting and helpful, together with a few brief comments. Most of these books have good bibliographies for use in specialist studies.

1 Introduction

General introductions to the science, technology and society field are: Rose, H. and S. (1970), *Science and Society* (Harmondsworth: Penguin); and J. Ziman (1976), *The Force of Knowledge* (Cambridge: Cambridge University Press), written mainly with physicists in mind, is enjoyable and has good illustrations. A more academic work is Spiegel-Rösing, I. and Price, D. de S. (eds.) (1977), *Science, Technology and Society: A Cross Disciplinary Perspective* (London and Beverly Hills: Sage), which has extensive bibliographies and specialist contributions. The *New Scientist* is an excellent source for news of developments in science and technology and their possible social implications. Kumar, K., (1978), *Prophecy and Progress: The Sociology of Industrial and Post-Industrial Society* (Harmondsworth: Penguin), ch. 6, gives a good review of the arguments in the literature on the post-industrial society debate; and Freeman, C. and Jahoda, M. (eds.) (1978), *World Futures: The Great Debate* (Oxford: Martin Robertson), reviews the main futurology studies up to 1978 in chs. 2 and 3. The journal *Futures* contains interesting material relating to recent scientific and technological developments.

Chapter 2   Historical Background

Two excellent general histories are Bernal, J.D. (1969), *Science in History*, vols. 1-4 (Harmondsworth: Penguin), which covers science and technology, and takes a strong 'externalist' approach; and Mason, S.F. (1956), *Main Currents of Scientific Thought* (London: Routledge & Kegan Paul; New York: Abelard-Schuman; later published as *A History of the Sciences)*. Various views on the scientific revolution can be found in Basalla, G. (ed.) (1968), *The Rise of Modern Science: Internal or External Factors: ?* (Lexington, Mass.: Heath). Landes, D.S. (1969), *The Unbound Prometheus: Technological Change and Industrial Development in Western Europe from 1750 to the Present* (London: Cambridge University Press), is a detailed and scholarly study. The wider aspects of the Industrial Revolution in Britain are discussed in Hobsbawm E. (1969), *Industry and Empire* (Harmondsworth: Penguin); and Taylor, P.A.M. (ed.) (1958), *The Industrial Revolution in Britain: Triumph or Disaster?* (Lexington, Mass.: Heath). Carr, E.H. (1962), *What is History?* (Harmondsworth: Penguin) deals with general questions of historical explanations.

Chapter 3   Philosophy and Sociology of Science

Some of the material in this chapter draws heavily on Richardson, M. and Boyle, C. (1979), *What is Science?* (Hatfield: Association for Science Education). Chalmers, A.F. (1980), *What is this Thing called Science?* (Milton Keynes: Open University Press), provides an excellent introduction to the philosophy of science. Kuhn, T.S. (1970), *The Structure of Scientific Revolutions* (Chicago and London: University of Chicago Press), is also strongly recommended; as is Ravetz, J.R. (1971), *Scientific Knowledge and its Social Problems.* (Oxford: Clarendon Press). The Popper–Kuhn debate is covered in the papers in Lakatos, I., and Musgrove, A. (1970), *Criticism and the Growth of Knowledge* (London: Cambridge University Press). Excellent collections of recent writings on the sociology of science are to be found in Barnes, B. and Edge, D. (1982), *Science in Context* (Milton Keynes: Open University Press); and Knorr-Cetina, K.D. and Mulkay, M. (1983), *Science Observed* (London and Beverly Hills: Sage).

Chapter 4 Politics of Science and Technology

Williams, R. (1971), *Politics and Technology* (London: Macmillan), provides a clear exposition of many of the most important issues in the field. Salomon, J.J. (1973), *Science and Politics* (London: Macmillan), considers in greater depth the political aspects of the social relations of science. Ellul, J. (1965), *The Technological Society* (London: Cape); and Marcuse, H. (1964) *One-Dimensional Man* (London: Routledge & Kegan Paul), consider 'scientism' as a philosophy of industrial societies. Gummett, P. (1980), *Scientists in Whitehall* (Manchester: Manchester University Press), gives a good account of the advisory role of British scientists in government.

Chapter 5 Economic Perspectives

For a very readable historical introduction to economic thought, covering in more detail the ideas of the economists discussed in this chapter, the student is referred to Barber, W.J. (1981), *A History of Economic Thought* (Harmondsworth: Penguin). For those who wish to continue an investigation into the development of economic ideas at a more advanced level, a highly recommended comprehensive text is Blaug, M. (1977), *Economic Theory in Retrospect* (London: Heinemann). For a treatment of the problems of environmental pollution, a useful introduction is Kneese, A.V. (1977), *Economics and the Environment* (Harmondsworth: Penguin), a well set-out book which deals with the economic aspects of the environmental costs of pollution with many examples. For a thought-provoking book, see Hirsch, F. (1978), *Social Limits to Growth* (London: Routledge & Kegan Paul).

Chapter 6 Food and Agriculture

Tannahill, R. (1975), *Food in History* (London: Paladin), provides an interesting description of developments in food and agriculture through the ages. Carson. R. (1962), *Silent Spring* (Harmondsworth: Penguin), although written over twenty years ago, is still a stimulating book for the reader considering the ecological ill-effects of food and agricultural technology. The economic aspects of farm policy in industrialised nations is surveyed in James, G. (1979), *Agricultural Policy in Wealthy Countries* (London and Sydney: Angus & Robertson). George, S. (1976), *How the Other Half Dies* (Harmondsworth: Penguin), provides a moving critique of the food problems of the Third World. For readers interested in nutritional aspects of food processing, Bender, A. (1978), *Food Processing and Nutrition* (London: Academic Press), provides a good introduction to this specialised field. An account of developments in genetic engineering and biotechnology, of relevance to this and the next chapter, is Yoxen, E. (1983), *The Gene Business* (London: Pan).

Chapter 7   Health and Medicine

McKeown, T. (1979), *The Role of Medicine* (Oxford: Basil Blackwell), gives case studies of the role played by medical science in eradicating various diseases and improving health. Kennedy, I. (1981), *The Unmasking of Medicine* (London: Allen & Unwin), gives a simply written summary of most of the key ethical issues in health and medicine today. Ledderman, E.K. (1970), *Philosophy and Medicine* (London: Tavistock), provides a clear exposition of the philosophical areas of medical science. Klass, A. (1975), *There's Gold in Them Thar Pills* (Harmondsworth: Penguin), supplies a critique of the use of drugs in western society, and of the activities of the drug industry in particular. Norton, A. (1973), *Drugs, Science and Society* (Glasgow: Fontana), provides a balanced account of the use of drugs and their social effects. Doyal, L. (1981), *The Political Economy of Health* (London: Pluto), is a critical study of the practice of medicine, and includes a commentary on the problems of the Third World.

Chapter 8   Energy

Foley, G. *et al.* (1981), *The Energy Question* (Harmondsworth: Penguin), is an excellent introductory book, covering many aspects of energy. In a different style is Wild, M. (1980), *Energy in the 80s* (London: Longman). More technical discussions will be found in Scientific American (1979), *Energy* (San Francisco: Freeman); and Patterson, W.C. (1977), *Nuclear Power* (Harmondsworth:

Penguin). A readable book on oil is Odell, P. (1981), *Oil and World Power* (Harmondsworth: Penguin).

Chapter 9 Military Technology

A simple introduction to modern weapons is given in Cox, J. (1977), *Overkill; The Story of Modern Weapons* (Harmondsworth: Penguin). For an introduction to the historical background, Fuller, J.F.C. (1972), *The Conduct of War 1789–1961* (London: Eyre Methuen), could be consulted. All of the SIPRI *Yearbooks* give a mass of detailed information and discussion on almost every aspect of modern military technology: SIPRI (1983) *World Armaments and Disarmament: 1983 Yearbook* (London: Taylor & Francis), is a good example.

Chapter 10 Telecommunications and the Mass Media

A set of readings which provides an introductory discussion is McQuail, D. (ed.) (1972), *Sociology of Mass Communications* (Harmondsworth: Penguin); which can be usefully supplemented by Benjamin, G. (ed.) (1982), *The Communications Revolution in Politics* (New York: Proceedings of the Academy of Political Science), which concentrates on the political aspects of mass communications. Murphy, B. (1983), *The World Wired Up: Unscrambling the New Communications Puzzle* (London: Comedia Publishing Group), gives a good account of the way in which different countries are handling the new developments in telecommunications. Also of interest are Lord Windelsham (1980), *Broadcasting in a Free Society* (Oxford: Basil Blackwell); and Wenham, B. (ed.) (1982), *The Third Age of Broadcasting* (London: Faber & Faber). Sampson, A. (1973), *The Sovereign State of ITT* (London: Hodder & Stoughton), gives a readable account of the growth of a major multinational company in the telecommunications industry.

Chapter 11 Scientific Management and Work

A good starting point is Braverman, H. (1974), *Labor and Monopoly Capital: The Degradation of Work in the Twentieth*

Century, (New York: Monthly Review Press). This could be followed by Wood, S. (ed.) (1982), *The Degradation of Work?* (London: Hutchinson); or Littler, C.R. (1982), *The Development of the Labour Process in Capitalist Societies* (London: Heinemann). Goldthorpe, J.H. *et al.* (1968), *The Affluent Worker: Industrial Attitudes and Behaviour* (Cambridge: Cambridge University Press), provides a thoughtful sociological account of the role of the worker in industrial society. Blackaby, F. (1979), *Deindustrialisation* (London: Heinemann), reviews the changing structure of industrial society and the problem of deindustrialisation. Mensch, G. (1979), *Stalemate in Technology* (Cambridge, Mass.: Bollinger), is a thought-provoking book analysing the economic problems of inflation and stagnation in industrial societies, and looking at the kinds of futures available to industrialised nations.

Chapter 12 and 13   Questions of Control

An excellent discussion of the subject of the control of technology, and many suggestions for further reading, can be obtained from the various units written for the Open University, Course T361 (1978), *Control of Technology* (Milton Keynes: Open University Press); and the associated course reader, Boyle, G., Elliott, D.A. and Roy, R. (eds.) (1977), *The Politics of Technology* (London: Longman/ Open University Press).

Name Index

Subject Index